Road Trip: Huautla
The Mushroom Cult

Ralph Trout

Copyright © 2018 Ralph Trout
All rights reserved.

The author retains the rights to all photos and text contained within this book.
Nothing may be used without express written permission.

ISBN: 978-1-64440-736-3

Dedicated to my lifelong friend Bill Swazuk,
fervently committed to keeping life interesting.

Contents

4 Guiding Light

9 Tampico

16 Mexico City

20 Pyramids

23 The Basilica

26 Bull Fight

27 Mt. Popocatepetl

31 Reservoir Camping

33 Veracruz

36 Yucatan

47 Oaxaca

49 On the Road to Huautla

61 Bill and Mike's Trip

65 New Experiences

73 Main Street Huautla

82 Van Mechanics

103 Maria Sabina

110 Psilocybin

FOREWORD

I toured the US throughout 2017 hopeful to discover at least a few of the causes of the current declensions, divisions, and deceptions. I recalled the early 70's. Those were similar times and in 1973 I decided to make a road trip to Mexico along with my friend, Bill. This is that story.

Mexico translates to adventure, remarkable pre-Columbian Indian cultures, and Shaman magic. That's what we were looking for back in 1973. Coincidences and good luck fulfilled our hunger more than we ever expected. We never searched for what we found, and got answers to questions we weren't smart enough to ask. Dumb luck and an old Indian woman in the crest of the Sierra Madre Mountains delivered a unique adventure. Fifty years later and after considerable research, we realized the woman we'd stumbled upon had been integral in changing the modern culture of the world. That is quite a statement, but María Sabina and the sacred hongos opened doors that evolved perceptions. In turn, drastically alternate lifestyles were created. Few know anything about her. This petite, weathered, Indian grandmother in the absolute middle of nowhere initiated the psychedelic revolution that eventually turned on the world.

Magic mushrooms are timeless. Ancient peoples used them as a tool for spiritual inspirations while being guided by the tribal Shaman. The Mayans consumed psilocybin mushrooms in religious ceremonies. Their descendants still enjoy them for mystical purposes. Stone carvings of mushrooms growing out of the heads of Mesoamerican Indians dating from 1000 BCE have been found in diverse locations from graves in Southern Mexico to the Guatemalan Highlands.

The Aztecs named the special mushroom 'teonanácatl' (tone-un-a-cAT-TLE) and enjoyed the effects during festivals. In the Aztec language of Nahuatl, 'teonanácatl' translates as - teó (blessed) and nanácatl (mushroom). The Catholic Franciscan monks described the mushroom rituals they encountered among the Aztec culture in the 1530s. The Aztecs ate the fungi with chocolate and honey. The ritual was termed monanacahuia or 'to become a mushroom'. The Aztecs attributed the spiritual effects of the magic mushrooms to a specific god, Xochipilli (sho-chy–PEE), the Prince of Flowers. The Indians believed mushrooms to be special flowers. It took the invention of the microscope to differentiate fungi from plants. Mushrooms were not cultivated until 1750.

The 'shroom ritual recorded by a Christian missionary depicted shamans and elders searching for the sacred mushrooms only after a fast and prayer. They picked at daybreak when the morning breeze signified it was permitted by their gods. The belief then was that these fungi are a gift from the Gods, sent on the morning wind and growing without seeds.

After Cortés conquered the Aztecs, Catholic missionaries —especially Franciscan monks —almost succeeded in eradicating all the Pre-Columbian religious cultures, including the worship of magic mushrooms. It was the belief of the Catholics that mushrooms were fiendish implements of evil. In their place they installed the Holy Communion with bread and wine. Even though converted to Christianity, Indians in the original mountainous cultivation areas secretly continued the religious and medicinal use of their sacred mushroom.

The Mazatec tribe is the most known for its consumption of magic mushrooms. The Mazatecs have various names for the mushrooms, such as, 'nti sit tho' translated as 'little ones who leap forth' and 'ninos santos' meaning 'holy children'. This mountain-dwelling tribe, known as the 'People of the Deer' and mainly composed of farmers, has been able to quietly exist for countless centuries in the Sierra Madre Mountains of southern Mexico. The majority of the mountain people speak their native language as well as Spanish. The Indians' lives are still very basic and they live in union with nature depending on good soil and pleasant weather with enough rain to cultivate many subsistence crops, including coffee.

These Indians exist in conjunction with the earth. They live in thatched adobe houses, cook over an open fire, and work their gardens with hoes and machetes. Dirt paths wind through the hills to their groves of coffee trees for walkers and those lucky enough to have a burro. As the mountain ranges converge, so do the eras in which they live. They speak their native language from the distant past and live close to nature, but they are influenced by the modern society of roads, transport, and electronic communications of 21st century Mexico.

The mountain farmers are Roman Catholic, but many have fused it with the local shaman deities. In each village, the springs, waterfalls, and caves are still held to be sacred —each believed to have a living spirit. The mushrooms enable the Mazatecs to connect and communicate with the various spirits of their immediate existence. With the assistance of these spirits, they believe diseases can be cured and the future foreseen. The spirits can take them to meet their god. The sacred mushroom is the Mazatecs' teacher, doctor, and psychiatrist.

Huautla (watt-la) de Jimenez lies deep within the mountains of Oaxaca (wa-HAK-a), a tedious six-hour drive from the nearest city, Teotitlán. The town is home to twenty thousand Indians and Mexicans. During the Aztec Empire, Huautla was the largest village within the Mazatec Region, and they paid tribute, perhaps with the region's sacred mushrooms. The Life article pushed psychologist Timothy Leary to first experience psychedelia in Mexico in1960. After his 'awakening', Leary and Richard Alpert initiated the Psilocybin Project at Harvard. After tuning in and turning on, they were

turned out and fired from Harvard in 1963. The floodgates had opened —
and by the 1970's while a few people were heading south to gain a 'shroom
experience, wild and cultivated psilocybin mushrooms became abundantly
available in North America and Europe.

In 1965, Dr. Timothy Leary commented that he had: "learned more
about my brain and its possibilities… and more about psychology in the
five hours after taking these mushrooms… than in the preceding 15 years of
studying and doing research in psychology."

The magic mushroom has come a long way from the mountains of
Mexico. John Hopkins University conducted a study in 2006 on the spiritual effects of psilocybin. Thirty-six adults tried the drug under laboratory
conditions. Twenty-one reported having an unforgettable, intense, mystical
experience. "Most of the volunteers looked back on their experience up to
14 months later and rated it as the most - or one of the five most - personally meaningful and spiritually significant experience of their lives," according
to lead investigator Roland Griffiths, Ph.D.

I kept a diary during our trip to Mexico and snapped a lot of photos.
The experience with María couldn't be forgotten. Writing this account in
retrospect, research divulged more interesting information on many of the
locations we visited.

You will carry that thought as your life's question;
is what I speak true, or created to confuse?

GUIDING LIGHT

It was a '70s summer and you had to have lived through the era to have enjoyed and perhaps understand it. Like Dickens: 'It was the best of times and the worst of times'. We were free spirits—loving life with few worries. The Army draft was behind us, lucky in the lottery. The American Dream dangling promised fortunes pulled us forward. The worst of times included the shady Nixon crew and the ever-present military industrial complex that Ike had warned us about.

The summer of '73 was about questions, where to go and what to do. Was it time to start maturing into an adult? Meanwhile, our usual answer was 'Let's take a pause for the cause and get high'. Our questions centered on American values. Not much has changed in almost fifty years. They were our questions—because my buddy Bill and I thought alike. We were true capitalists. The plan, drummed into our heads, was to beat the system before it beat us. Suddenly, everything was political: our attire, our hair, and our music. We were young—fresh out of school, but not yet in debt, indoctrinated into the cash-only system. Credit cards were rare and banks actually paid interest. This was prime time for American Express Travelers' Checks.

Two others joined for a short two-week, south-of-the-border vacation. Sam had graduated from Pitt with me and was returning to get his master's degree. This was his chance to let loose. Earl had been a Marine sergeant in Vietnam and wanted/needed to relax. This was before Post-Traumatic Stress Disorder had been labeled. He'd been a hometown friend who returned with the thousand-mile vacant stare, garnished with an extremely high IQ. He had seen some shit, but never talked about it.

The Travelers — Ralph on the left and Bill

I'd had a year of Latin and another of introductory French in high school and barely got through college French—but no Spanish. Sam was the only one versed in the language of our destination. We purchased English/Spanish dictionaries and a copy of Fodor's Guide to Mexico.

Who could say where our society was heading? The system was moving so quickly to either stamp out every new cultural wave or profit from it. Everything was about the time. Our parents' and grandparents' time, having fought the great wars when our country, culture, ideals, and specifically freedoms were attacked; that time was totally different from our reality. Those conflicts were to save our liberties. The Vietnam War wasn't about anything except sending young men into a jungle to fight and die, pretending to stop the spread of communism. The war kept expanding and the news reported more than any government agency wanted. Shady, cross-border, unauthorized conflicts were becoming the norm for an increasingly spooky, unrepresentative government.

Going south of the border, our object was to gain perspective. The '70s were the beginning of the technological onslaught, with great changes that occurred almost daily. Continental America was timeless, yet almost everything pre-Columbian had been paved over. The true, geographic ancestors had been transformed into rebel heathens fighting for their rights against the Great White Men and the European culture. A convenient history of America had been created. To the south, Mexico beckoned with huge monuments to a pre-Christian society, jungles, undeveloped beaches, and adventure.

America had become more of a paradox than a paradise. Every aspect of our lives was bought and sold. It was the initial era of psychological conditioning or programming. Orwell's 1984 with Big Brother still loomed in the future. Mexico was appealing. Perhaps that vast country still held a paradise —or at least some answers —even vistas that could provide a new perspective. Maybe there would be a small, isolated place that hadn't been touched by technology. The scene of a farmer with sombrero tilted, snoozing against a cactus was enough motivation.

A stroke of fate brought me an already classic '65 Volkswagen van. Not just any van, but a twenty-two window camper with a full sunroof and a luggage carrier. New radial tires and a new engine, I saw it advertised and bought it for $550. That was incredibly reasonable. It was in almost-perfect condition. The owner was a doctor who had bought it in Europe and did the tour, keeping every maintenance receipt. The blue and white van was a beauty to behold. It had a Blaupunkt AM/FM, with the unique search feature. We needed tunes in the days before cassettes, CDs, or MP3s.

The VW never gained a nickname beyond 'Volks'. A factory rebuilt 50-horsepower engine and new Sears Dynaglass radials were a blessing. It didn't have a kitchen, but all the cushions fit together and made a long, wide, very comfortable bed that looked up through a sunroof. With an ice chest and one of those green Coleman gas stoves, we were good to go. Under the seats was plenty of storage for spares and oil. If ever there were a vehicle that defined a life-style, it was the VW transporter. If we weren't hippies, at least we looked the part.

We drove straight south from Pittsburgh to Laredo, Texas and crossed the border into Nuevo Laredo, Mexico without incident. After only two hours, AAA supplied the special car insurance necessary and we got our traveling papers. Once in Mexico, it was obvious the officials wanted US dollars. Our VW van could have been loaded with weapons for an insurrection —yet the single dollar we tipped the border guard ended any search.

The first vision of Mexico after the raucous border town was the small village of Cuidad Mier. My diary reports it was a small, quaint, rural town. That was then. Timing is everything, because in February 2011 that quaint town exploded in a gang war that began between the Gulf Cartel and the Los Zeta gang. The Zeta's were deserters from the Special Forces branch of the Mexican Army. Drugs weren't just for recreation anymore.

Our dream was to relax and explore some deserted beaches. The quickest route to the ocean was by way of Gulf Rancho La Carbonera. The Mexican road trip took an almost immediate detour. Unknown to us, and unfortunate, torrential rains had wiped out the road. We were forced to cross back into the US at McAllen, Texas. Of course, a VW bus coming out of Mexico after only two hours had to be searched. We could have made our rendezvous, stashed the contraband, and been ready to return to make some cash. That wasn't our initial intention. Of course, we liked our herb, but the plan was to relax—to find a natural high.

The customs officials did a comprehensive search, slitting every seat cushion before bringing in the dogs to prove we were clean. Nothing had to be repaired or replaced as they were protecting a drug-free America. It's obvious how that worked out.

Again in the US, we headed south on another route, crossing at Reynosa. The US officials permitted us to pass, but the Mexicans decided we were now undesirable. It was the late shift and they required compensation to enter. A short, thin uniform offered us the required visa for $10 each. In a scant few hours, we had been blocked by impassable terrain, unloaded and re-loaded the van with three months' worth of supplies, and been strip searched. To cross the border, we crossed his palm, and our journey continued, now legal.

Pesos were needed and the local banco obliged. We again realized border towns suck. After asking several people, we finally got directions to follow Highway 97 towards San Fernando, still in Tamaulipas state. Bill nodded under the open sunroof and got fairly toasted. He wasn't in any shape to enjoy the first beach at Playa La Carbonera.

La Carbonera Beach

 Our journey continued to be bolloxed by the roads. The earlier storm had washed out another along our path. The van had to be pushed through a few mud pits. Our goal was a ranch; getting there required using a floating ferry, pulled by horses, across a swollen stream. The road worsened and we had to turn around, lucky to catch the ferry before it closed for the night. Back in San Fernando, we consumed local frijoles and Dos Equis beer for the first time and slept well with full bellies. This had been a steep learning curve on Mexican road conditions.

 The vacation really began at Cuidad Victoria, a tidy, authentic Mexican town wrapped around an impressive church. Victoria was big enough to have a supermarket and its shelves revealed how much influence the US had on the local economy. Gregory, a sleazy, supposed guide became the first of many to offer 'drogas', especially pills. Courteously refusing, we exited, driving across a bridge to catch the first glimpse of the basic river laundry, with women beating the dirt out of clothes on rocks. They were not excited about getting their photos snapped. A bit farther downstream, the VW bus got a cleansing and we got our first swim. A rule to follow was quickly learned: never swim downstream from a village.

 We parked next to the stream, prepared a good meal, and plotted our course south to Tampico. Our path wound through scenic foothills labeled with tended, white crosses where other travelers had perished. Mesas rose as the bus rambled from the arid plains to what we misinterpreted as jungle—but was only succulent land filled with mango trees and sugar cane farms.

ROAD TRIP: HUAUTLA

Bill Cooking

At the small village of Mante, the first bit of 'gringo go home' hit when a local girl called us a variety of bad names —assuming correctly that we were hitting on her. Out on a lonely road we decided to call it a day. Every evening was a series of coin flips to decide who slept on the van's cushions and who got the front seat. Even three was a bit crowded in the rear, but usually we were exhausted or beered out and it didn't make much difference.

That evening, Sam and I set up our jungle hammocks. These were Nam surplus with a sturdy canvas bottom, and you zipped yourself into a very strong mosquito netting with a water-repellant, nylon roof. It was ingenious as it could be staked on the ground as a tent or strung between two trees. Sam's hammock decided it would rather be a tent, as he hit the ground and concluded the van's front seat would be a better choice. An early morning rain was my wake-up call.

TAMPICO

The altered route led to Tampico, with sunshine on a beautiful beach bordered by the oil industry. We were equipped, ready, and eager to camp—yet there were no campgrounds along this stretch. The guide-book served up cold-water-only apartments for $2 a double at the Ritz Hotel on the beach. Everything inside and out was painted the same pale green, but it was a luxury to sleep in a bed again. Food was good and inexpensive. Tacos and tostadas and sweet Coca-Colas or ice cold beers were reasonable. The dollar was strong. Sometimes lunch, dinner, and brews would cost only pocket change. Of course, we considered bottled beverages were our first line defense against 'Montezuma's Revenge'. However, we sadly learned the combination of Carta Blanca beer and cheap tequila erupted another version of Mexican revenge.

Our first south-of-the-border turn on came from two Mexican fishermen, Ernesto and Pepe. After a few cervezas, they brought out some of their country's best smoke, dubbed 'monkey tails'. We had never heard the term before, but soon discovered the flowers were so resinous they held together as you squeezed out the few seeds. This is before the time of 'sensimilla', 'clones', 'skunk', and all the other terms that have become so common. High Times magazine wasn't yet on the shelves. We were not very weed-educated and only accustomed to smoking 'around-town brown'.

Ernesto, between long tokes, explained the best smoke was grown in Oaxaca–(wha-HACK-a). Hell, it took days before we could pronounce the Mexican state. Oaxaca sounded oriental—perhaps Grateful Dead-ish—but that's where the sailors swore the best pot was grown. We were from Pittsburgh, the Steel City, then strangling in smog. Our home location attracted very few exotic smokes. Michoacán was famous for long flower tops—and of course, we knew of the fabled Acapulco Gold, but where the hell was Oaxaca?

Ernesto professed to be an expert herbalist and knew his country. Marijuana was termed 'mota'. Oaxaca was a special place, he proclaimed. Not only Grade A mota, but magic mushrooms. Hippies roamed the mountain pastures eating crazy mushrooms. Our reply was, 'Bull-shit,' assuming he exaggerated to impress us. Maybe we had long hair, embraced free love, and were anti-war, but we weren't that naïve to believe magic mushrooms were abundant in this far off Mexican state of Oaxaca, free for the picking.

This journey wasn't about getting buzzed, but more about anthropology and discerning the evident differences in our neighbor's culture. Imbibing, soaking, and perhaps smoking might be a track to enjoy and define the variances. At the least, it sounded like a plausible excuse.

Pepe said 'hongos'—Spanish for mushrooms,—grew everywhere in

Oaxaca and many gringo hippies traveled to graze. The real Mexican drug scene they described was paranoia. According to their tale, every Indian and sailor gets high. The Indians grew it; the sailors bought and enjoyed it. Few older Mexicans smoked dope, but most had tried it. Locals would probably be willing to trade grass for papers, as rolling papers were illegal. That was good to know because we'd brought a couple of packs. The usual way to roll a joint was to carefully and tediously scrape the foil from the paper inside a cigarette pack. Pipes weren't a Mexican thing.

Wild Magic Mushrooms

Speaking of Mexican things, Bill was the first to succumb to the shits, followed hours later by Earl. They were exuberant to share the event, doubled over with cramps and searching out the nearest baño or secluded spot. Common sense should have implied that all the veggies we were consuming in ensaladas were washed in the same bad water. It may have been a good thing that the veggies added some fiber.

Some of us had clenched cheeks, but we found the local tourist office to get some better info on Tampico. It was there we met Ed, who was a grad student from Cleveland doing research on the Mexican guerillas. He was a straight-laced, yet ballsy guy—ready to trek into the mountains to locate rebels evading the local army. It would be one hell of a thesis if he survived. Undoubtedly, some of the not-too-idealistic rebels evolved into well-armed drug cartels.

The fishermen, Pepe and Ernesto, said there was a possibility we could crew on shrimp boats and get our fill of fresh seafood while getting a tan. After the tourist info, the bus headed to the harbor. With limited española, we learned none of the boat's masters were around and mañana would be better. We learned, with everything, mañana was always better.

Some kids were tossing hand nets and bringing in small fish. A small beach bar fried them in hot oil. They were the freshest and some of the best tasting fish ever. The afternoon was accented by a beach bar brawl between two mature Mexican brothers fighting over a couple of pesos. Existence was noticeably inexpensive, but not cheap.

Sam and Earl decided there wasn't enough time for a voyage on a shrimper and they were both already sunburned. Mexico City would be next. Luckily we met two men who happened to be driving that way. Directions were difficult to understand, between turn right—derecho—and go straight ahead—directo—rather than al frente. Unfortunately, it did not quickly occur to us that people seldom traveled from their villages. When they did, it was on a bus, so they never paid much attention to turns. Truck drivers were the best information sources, but they were not often sociable.

The drive from the coast into the mountains was more scenic than we could have imagined. Lush, green, misty overlooks were frequent along the fairly good road to the town of Tamazunchale (Tam-UXUM-tzalle). Again, dumb luck prevailed, meeting a local who not only spoke English but had learned it in Ligonier, Pennsylvania, about fifty miles south of Pittsburgh. He worked at the Hotel Mirador, the best in town. Our new friend got us two rooms for less than five dollars a night. Chasing the long day of driving and the chill from sunburn combined with a much lower temperature, a hot shower was a godsend.

Our path to the capitol meant more twists and turns, as the landscape transformed into an arid, high plateau. Stopping for some food, our language lessons continued as we learned comida corrida was the slang for a full course meal. By then, Montezuma had been quieted with a combination of eating handfuls of uncooked rice and a few Lomotil. The former was suggested in a market and the latter in the pharmacy.

Never fearful of different foods, we learned how to carefully pick and prepare purple cactus fruit: prickly-pears. We'd seen Indian women selling them along the streets and tried a few. Parked for the evening on a local football/soccer field, an Indian boy showed us the trick to using a rag so we didn't get stuck with needles and then paring the fruit with a knife.

Tula, the Toltec archaeological site outside of Hidalgo, was our first dose of Mesoamerican ruins. This is what we had traveled to view. The Toltecs were one of the pre-Columbian civilizations that had risen and fallen before the European onslaught. Tula was probably the best Toltec site, populated for two centuries from 900 CE. The center is a small pyramid carved with images our guide book told us were its rulers. The Toltec supposedly came to Tula (Tollan) after the demise of the huge pyramid city of Teotihuacán near Mexico City. Tula is believed to have had a population of sixty thousand—who for some reason, suddenly moved seven hundred miles south to what had been the Mayan city of Chichén Itzá. All the reasons for movements of these highly evolved Amerindian tribes are speculation.

Once the Toltecs re-inhabited Chichén Itzá, they built stone edifices in their own style.

This archaeological site was everything and more I'd hoped to see. It was the name of the statues that was most attractive, Los Atlantes de Tula, 'The Giants of Tula'. Atlantis, the Lost Continent, had always interested me. However, I learned the statues aren't named after Atlantis, but Atlas—probably because the statues probably held up a roof.

On top of the pyramid are four stone (not stoned, which became our common joke) figures twice our height. One of the guides for a nearby group (that we were too cheap to join) said they were of the great god Quetzalcóatl (KE-tso-co-ottle), or a king, or someone else named after him. The guide was a young lady from State College, Pa. working for college credits for Penn State. She told us the place had been discovered in 1940 after looters dug a hole. In 1973, most of the site was still unexcavated at that time due to lack of funds.

Tula Pyramid: Atlantes

These four statues are thought to be Quetzalcóatl and his brothers, each depicting cardinal points of the compass. Each statue is carved from four blocks, with the lower section representing legs and feet, the next two making up the mid-section, and the fourth the head and helmet. Quetzalcóatl, in some cultures, was considered the 'White Morning Star', the god of the West. He had all of the good guy qualities of light, justice, and mercy. He was also the god of the wind—and when thoroughly pissed off at man, threw a hurricane. The South was ruled by the Blue Huitzilopochtli (wee-tsee-loh-POCH-tlee),—the mean war god. The East was guarded by Xipe Totec (SHE-pay TOH-tek), who influenced gold, farming, and springtime.

The North was governed by Tezcatlipoca (tes-kaht-li-POH-kah),—the exact opposite of Quetzalcóatl. Tezcatlipoca is the god of conflict, the night, trickery, magic, and the planet Earth. The four immense statues are definitely warriors: adorned with breastplates, shields on their backs, feathered headdresses, and carrying flint knives and spears.

Our guide shared that Quetzalcóatl wasn't the fabled white god the Aztecs supposedly confused for Cortés, but the winged or plumed serpent—who not only brought knowledge that helped the tribes evolve from the Stone Age into great civilizations, but knowledge that could ultimately consume humanity. (Not far from a definition of modern technology.) In the early '70s, there were few Ancient Alien theorists. Erich von Däniken's Chariots of the Gods was published in '68.

The north side of the temple had carvings of snakes eating people. On the other sides were carvings of jaguars and coyotes, with eagles and vultures eating human hearts. Quetzalcóatl was depicted emerging from a creature that is the combination of a serpent, a jaguar, and an eagle. It was then our friendly guide winked and whispered the guys who did the carvings were probably doing 'shrooms.

The Penn State girl recited various myths and legends about Quetzalcóatl actually being the priest-king of Tula. One account had his rival brother, Tezcatlipoca,—'The God of the Night,' forcing Quetzalcóatl and his followers out of Tula around 1000 CE. (The Penn State girl was excellent at pronouncing all these names. Tëzca means 'mirror'; tl means 'fire;' and pōca is the resulting 'smoke'. So, the evil brother is the smoky mirror—The God of the Night—probably the evil smoking volcano.) Quetzalcóatl then wandered the east coast of the Gulf of Mexico, where, in one version, he sacrificed himself on a burning pyre and became Venus, the Morning Star. In another ending, Quetzalcóatl built a raft from snakes and vanished into the east. That's how the invader Cortés comes into these legends.

She continued with another version that had Quetzalcóatl arriving on Earth to rule the world as an ancient alien god. In this version, he created man from his own flesh. Cloning hadn't yet been heard of in the '70s. Her story evolved with a blush, as she continued, mentioning that he'd sliced his penis and traveled all over Mexico dripping blood. Again, she referenced 'shrooms—everywhere a drop of the God's blood hit the ground, magic mushrooms grew. After spawning the human race and 'shrooms, he became the obvious leader. Through his knowledge and assistance he raised the Aztecs to rule Mexico.

Quetzalcóatl was kind and benevolent, teaching his people how to use teonanàcatl (tone-un-A-cattle), the Aztec name for magic mushrooms, which translates to 'flesh of the gods'. Quetzalcóatl had no vices and remained celibate. The 'shrooms were religious, but his evil brother, Tezcatlipoca, slipped him a drink spiked with super 'shrooms and the father of mankind got wrecked—breaking his vows and having sex with Tlazoteotl

(te-lazo–teo-til), the beautiful and wicked goddess of adultery and sin (a very hot slut puppy). At the end of the story, Quetzalcóatl again heads east in a boat. Before he sets sail, he promises to return and rule again, but the boat catches fire and he burned to death. As with many religions, he is born again. As the Winged Serpent, he rises into the sky and heads towards the sun—leaving the door open for the Spanish conquerors.

Toltec Ruins at Tula: Chacmool Statue

Our freebie guide continued citing that the belief in the Winged Serpent God unified the tribes of Mesoamerican Central America. She knew representatives of all the tribes had gathered at the Great Pyramid of Cholula in Puebla State as a sort of yearly religious journey. (We later visited this great pyramid and it is huge. Cholula translates to 'handmade mountain'. This pyramid is bigger than the Great Pyramid of Giza!) The Spanish invaders witnessed the reverence the tribes paid to Quetzalcóatl and noted this reverence kept all the tribes unified. That's probably why there were so many willing sacrifices. The Spaniards forced a wedge, breaking the union by splitting the alliance and using the tribes of Puebla and Oaxaca against the Aztecs. There had been conflict between the Aztecs and the other tribes for a century prior to the European invasion. The Spanish just reheated it.

With these tribes' help, the Spanish were able to conquer the Aztec capital, as well as all the other smaller tribes of the vast plains. They even subjugated the Guatemalans. This was a win-win as the Puebla and Oaxaca tribes made out like literal bandits. Years later, they helped the Spanish transport goods via the Manila galleons from the Pacific coast across Mexico. These tribes also assisted and protected European religious conversions. Sadly, faith in Quetzalcóatl meant the downfall of all pre-Columbian cultures.

As she spoke of sacrifices, we saw our first Chacmool statue—where humans were stretched and had their beating hearts ripped out. (In April of 2009, the bones of 24 children were found in Tula dating to the Toltec period (900-1180 CE)—indicating they may have been sacrificed.) The weather that day changed to gray and bleak. We decided there was too much juju around those ruins to spend the night and headed to Pachuca, the capital of Hidalgo.

The VW bus was losing power due to the altitude—and so were we. Pachuca was supposed to have steam baths. Seems the Toltecs were into steamy rooms. Disappointed the baths were closed, we also closed for the night in a hotel's parking lot.

ROAD TRIP: HUAUTLA

MEXICO CITY

The drive from the beautiful mountains that echoed Mexico's great tribal history into the valley of sprawling Mexico City develops a vision of its rough, polluted present. It was immediately evident that Mexico City revealed the attitudes and living conditions of the country's future.

The largest city in Mexico didn't even make the top ten most populated cities in 1970, with less than four million. Even without a great command of the language, driving in the four crowded lanes wasn't as bad as trying to negotiate Washington D.C. or New York. This is not to say that the traffic didn't have its frightening moments.

In '73, with only half of the present population of the second decade of the new millennium, everything was crowded. Pedestrians packed the streets—except where Indian women had spread their scant blankets to sell cactus fruit. The highways and streets were slow moving. The subway had throngs waiting to pack inside already crammed cars. At that time, Mexico City had been ranked the fifth most densely populated city in the world.

We found the climate was not sweltering hot, but the air was thick with pollution. Mexico City is at an elevation of 2,200 meters. Mountains and volcanoes that surround it stretch to 5,000 meters with only a northward opening. The stagnant air has nowhere to go. The wind patterns circulate within the valley and retain the pollution rather than push it out. The elevation causes the incomplete combustion of fossil fuels; smog wins. Vehicle exhaust, smoke, and bean farts have no way to escape.

The huge city is located in The Valley of Mexico, also known as 'The Valley of The Damned'. Nice name, huh? The valley had originally contained five lakes, and is also in the Trans-Mexican Volcanic Belt. The 'Damned' probably comes from the continued colonization of its inhabitants. First they were conquered by several Amerindian tribes before the Spanish. After the European conquest, came the American oil companies.

The Valley of Mexico has been populated for at least 20,000 years. When the Spanish arrived in 1518, it already had one of the planet's densest populations, with over one million people. Colonialism, violence, and disease reduced the population, but it was finally repacked with a million again by 1900.

Our Fodor's Guide was invaluable. We arrived on a Thursday and it took a while to catch our bearings. Mexico City was astonishingly huge. This was long before GPS. Reading maps in six-lane traffic was an exciting part of our trek. Finally, we located the Federal District, where every building was covered in ceramic tile. We found a safe place to park and a telephone to call home.

Mexico City began as a bummer. The Capitol Hotel was packed and we got a 'mañana'. Sam had the cramps and needed a baño, ASAP. Since there were no rooms at the hotel, Sam was turned away from their facilities. Baños and rooms were available at the Hotel Toledo, and our walk-about began. Bill and Sam decided to stay close to the porcelain fixtures, while Earl and I proceeded to get almost hopelessly lost. Luckily, some street minstrels, now known as Maríachis, drew a map for a few pesos.

According to our guide book's recommendation, we dined at one of the city's five best restaurants—enjoying an incredibly tender steak for two dollars. Street tacos were two cents, but somewhat intimidating with the fear of stomach disorders. Everywhere was busy, day and night. Bill's sweet tooth led us to many pastry shops. We discovered a combined passion for the local dessert: plates filled with fruit and ice cream.

The following days were spent reading Fodor's Guide to Mexico and touring The Presidential Palace, the National Cathedral, and an occasional pawn shop. The Merced Market was huge! It had been operating since the Europeans rolled in. The market had anything you wanted, mostly food and restaurants, with shop after shop selling the same clothes. Our entourage spiked the Mexican economy—buying rugs, shirts, and sandals. We never lacked for food and most complete meals in those days were about sixty cents. (In 1974, a Big Mac was sixty-five cents) A fabulous meal at a top-shelf restaurant didn't cost three dollars. The day ended at the top of the Latin American Tower. From the top, we could see how Mexico City spreads forever, seemingly without end. At the edge of our vision, the city kept going on into the smog.

The next day was the Artisans Market, a-for-tourists-only place where we were delighted to discover English was spoken. A business wizard must have designed it, as it is a maze of stalls that entraps you. There were over two hundred stalls with local handicrafts. Everything seemed overpriced compared to what we had already seen in other, smaller markets. These people refused to bargain and that took most, if not all, of the fun out of it. Walking from museum to museum in the historic district of Mexico City consumed the remainder of the day. The Museum of Modern Art was impressive, with most of the world famous artists represented.

The Museum of Anthropology was stunning. Huge stone sculptures were planted around the building. The Aztec Stone of the Sun grabbed us and the huge Olmec stone heads held our attention. On the ground floor, every ancient tribe or civilization was represented. Until you view it, the depth and detail of these exhibits are difficult to convey. The Museum is a shrine to all of the past and present cultures of Mexico. We'd only seen the ruins at Tula and we knew little about Mexican history—except that the Aztecs were ruling when conquered by the Spanish.

The museum's second floor displayed every currently existing tribe in dioramas that occupied entire rooms. Limited Spanish always detracted from the experience. One room had a very unusual scene of an old woman standing over a man kneeling on the floor. They were obviously Indians. When you pressed a button an eerie chant could be heard. Intent on deciphering the plaque, my trusty Spanish-English dictionary was opened. A very attractive young lady, also viewing the strange scene, offered to translate. Monique was French and tri-lingual and spawned another series of 'shroom coincidences.

Monique explained that the scene depicted a religious ceremony of a current Indian tribe's mushroom cult. It was amazing; this babe knew about Oaxaca and the magical hongos. She also knew a special permit was necessary to enter that tribe's region in Oaxaca.

She told me about herself. At twenty-two, Monique was traveling the Americas with only three hundred dollars. She had already toured the United States and was headed south to Argentina with a hundred remaining. Apparently she could really stretch a budget. Being very slender and attractive, an Audrey Hepburn lookalike, didn't hurt. Unfortunately, she'd been busted for pot by the Federales in Mazatlán. She'd bought a kilo to resell to some gringos when the police interrupted the transaction. Confined to the local jail, she had to pay a small fortune of forty dollars.

Monique was now traveling with California friends. She had that quality to quickly make friends. It was another weird coincidence that her new friends had visited the mushroom village on a prior trip to Mexico. This young woman had many more tales to tell and volunteered to continue the next day.

The next afternoon, Monique filled us in on everything her Californian friends had told her. They had visited the mushroom village named Huautla (What-la), deep in the southern chain of the Sierra Madres where the east and west ranges converge. Her friends mentioned it would be a long, miserable drive over ridiculous roads and we would definitely need a 'special' visa.

Not wanting to lose track of Monique, a meeting was scheduled for the next day to gain more info on the 'mushroom village'. Another coincidence happened when Sam realized he'd lost his visa and was desperate to return by air. While waiting at the immigration office, Monique appeared with her friends. Bill and I chatted with her as the California friends attempted to gain extensions on their visas. Perhaps due to their long hair and beards, they were disappointed—no more legal time south of the border.

Whispering in the rear corner of the vast immigration office, the Furry Freak Brothers look-a-likes regaled us with tales of a mushroom religion lying deep in the mountains. The place sounded a bit more like Shangri-La than present day Mexico. The two guys rambled on about how visiting tourists were being swamped by crazy Indians selling 'shrooms/hongos wrapped in banana leaves as soon as they entered the village. We found it difficult to believe any place would have a drug culture that wide open. The two freaks shrugged as they left. "Think what you want. Huautla is truly different. The only way to believe, is to visit."

Our suspicion grew to disillusionment after an immigration officer flatly stated our present visa had no restrictions on any location within Mexico. The Californians' story sounded like a tall tale intended to fester in our imaginations. We had no concrete facts concerning the mushroom village—but had seen the display at the museum. So, it had to be real—but on what scale? Was Huautla a small, hidden Indian cult that lurked in a far-off village, or was it really the full-tilt 'shroom boogie the freaks described?

It was Sam's last days south of the border. We found a great restaurant in the San Juan Le Tran. Filet mignon was only two bucks. Sipping a bottle of tequila until early in the morning, we traded tales of working in the steel mills.

On the last day of our expedition in the capital, we spied an exclusive book store in The Federal District. A recent issue of Time in English was the attraction. Bill always wanted to keep aware of current events in the US. Inside the book store, we noticed a set of books titled Los Indios des Mexico. They were voluminous. After leafing through a few, I discovered a section on magic and sorcery. This was when Carlos Castaneda's book, The Teachings of Don Juan, was a current bestseller. Neither I nor Bill had read any of his works—and still haven't—but we knew we were onto something.

Castaneda wrote of his training in the realm of Yaqui peyote magic under the direction of Don Juan Matus. Over the years, his tales of magic and sorcery filled twelve books. Were his stories completely fiction, aimed at drug fueled readers? Bill and I scratched our heads wondering what would happen where we were headed. Were the stories made up of truth combined with exaggerations or fiction? Perhaps we could confirm Huautla was a reality. Our vacation was at a point of decision. Sam was leaving; should we head directly to the reputed mountain Shangri-La?

Again, my limited Spanish held me at bay. Not being able to read the couple of hundred pages of Los Indios, the pictures had to suffice. Most of the articles portrayed cactus peyote ceremonies, and only a very few depicted the fabled mushrooms. This at least gave us an idea of the appearance of hongos so we could forget the fear of randomly eating poisonous toadstools.

The prime importance of this vacation wasn't to get high. Our 'need' to indulge receded the farther we drove south away from the chaotic US. Our aim was to enjoy beautiful days in exotic places, soak in pre-Columbian history, not force anything to happen, and not get into trouble. No matter how we played it, someone or something always popped up and pointed the way to the land of the magic mushroom.

Equipped with the knowledge from Monique and the California freaks, we set our sights on Huautla—but only after we had explored more scenery and interesting locales. Legendary Huautla became the pot of gold at the end of the rainbow, but we decided it had to remain at the end. It would be a tremendous let down to miss an entire country just to get high.

PYRAMIDS

First on our list of places to see were the huge pyramids at Teotihuacán (tee-oh -TIA-wah- kahn). Bill and I had consumed Erich von Däniken's Chariots of The Gods and we were believers. The pyramids were an absolute must-see. It should have been a short car jaunt about forty miles northeast of the city, but as usual, traffic was congested and it took two hot hours.

We were not ready for the immense scale of the Pyramids of the Sun and Moon, or the amount of other tourists. These pyramids were gigantic, some of the biggest structures in the Western Hemisphere.

The ancient metropolis had many pyramids aligned along a lengthy boulevard that had acquired the name 'Avenue of the Dead'. Looking at the straight road with our trusty compass, the smaller Pyramid of the Moon was located slightly east of north. Today, researchers have noticed a similarity between the Mexican and Egyptian pyramids: both sets of great pyramids seem to be identically aligned to depict the belt of the constellation Orion.

Years ago, when we rolled into this colossal Mesoamerican city, its construction was attributed to the Aztecs. That theory has since been revised. No one knows who or when these stone monuments were created. Teotihuacán is the name that was given to the site by the Aztecs and it means 'City of the Gods' or 'where men become gods'.

Teotihuacán is a huge site. The main avenue, The Avenue of the Dead, is more than a mile long. In the hot sun, it seemed like we plodded forever. The Aztecs supposedly gave the name to the road because the mounds on either side appeared to be tombs. The Pyramid of the Sun is huge, with a base that rivals the size of Khufu's Pyramid at Giza. But the Mexican version is less than half as tall as the Egyptian and faces the sunset.

Research for this book revealed that Dr. Vincent Malmstrom of Dartmouth College theorized the creators of Teotihuacán had planned the city with such precision that it was also aligned to Orizaba, the highest mountain. Since the Pyramid of the Sun looks to the sunset, Malmstrom equated that the city is orientated to August 13th —'the day the world began' according to the ancient Mesoamerican legends. Amazingly, the end of the Mesoamericans by the Europeans came when Tenochtitlan, the Aztec capital, was captured by Hernán Cortés on AUGUST 13, 1521. That may be predictive or coincidental, but none-the-less the sunset of a great culture.

One theory found researching these pyramids is they were constructed during the first century and over one million mud bricks were used to fill the interior. Excavations in 1971 found a tunnel under the Pyramid of the Sun that ended in multiple ritual chambers.

The second biggest Pyramid at Teotihuacán —the Pyramid of the Moon

—is located at the northern end of the Avenue of the Dead. The Pyramid of the Moon is twenty meters shorter, but is naturally elevated, so it appears to be the same height as the Pyramid of the Sun. Digs done over the years under the base near the staircase revealed a tomb with a male skeleton. In '73, few tombs had been found in Mexican pyramids, except for the pyramid at Palenque.

When Cortés conquered the Aztecs, he asked the natives who built the colossal city. The Aztecs replied that they were not the builders of Teotihuacán. According to their legends, the city was built by the Quinametzin, a race of giants who came from the heavens in the time of the Second Sun. Teotihuacán can be translated to the city 'where men become gods'.

The Aztecs named the pyramids of Teotihuacán, 'Tonatiuh' —'Home of the Sun and Metal', and 'Tzaquati' —'Home of the Moon' The Aztecs reported that when their people first inhabited the area, the Pyramid of the Sun was covered with black basalt. The Aztecs considered Teotihuacán a sacred site, even though it had been abandoned long before their time.

Today, the majority of archaeologists cannot explain many aspects of Teotihuacán. Researchers believe the ancient city reached a peak population of around 200,000 inhabitants, but due to unknown reasons, the city was abandoned between 650 and 750 CE. It is believed that the inhabitants of Teotihuacán were a mixture of primitive, native peoples of Mesoamerica. If this had been the case, they would have lacked the basic elements for the development of civilization and writing. This leads to the greatest mystery: how a primitive people obtained the extremely high mathematical and astronomical knowledge necessary for the pyramids' construction and alignment.

Another surprising discovery was made in the early twentieth century when researchers exposed a thick layer of mica on top of the Pyramid of the Sun. Sheet mica is now used in electrical components, aerospace components, and radar systems. Mica has not been found in any other archaeological site in the vicinity, or anywhere in the Americas. The type of mica used in Teotihuacán is believed to have originated in Brazil. What was the purpose of dedicating all the labor to locate, transport, and install the mica nearly two thousand years ago? Inquiring minds want to know!

Another mystery: the Pyramid of the Sun was covered by a layer of 4 meters of dirt —which archaeologists believe was placed there on purpose. Researchers needed five years to remove thousands of tons of dirt to clean the base and surface of the Pyramid. Was the mica used as a beacon and the dirt to silence it?

Due to the many carvings, the third pyramid is supposedly dedicated to Quetzalcóatl. He was the main god of the Aztec priests; the god of learning, knowledge, and the wind. Teotihuacán was the first culture to use the symbol of a feathered serpent as an important religious and political symbol. This temple has prominent feathered serpent figures with alternate serpent heads. The earliest depictions of Quetzalcóatl as a feathered serpent god are that of an actual snake. The Plumed Serpent, an ideogram of the Quetzalcóatl, alludes to one of the most powerful forces of nature: wind.

More than two hundred sacrificial burials have been located at this pyramid, probably as part of the dedication of the temple. There are burials of both men and women, grouped in different locations. The males, accompanied by the remains of weapons, lead to the conclusions that they were probably warriors in service to Teotihuacán, rather than captives from opposing armies.

Likely due the current wrath of Quetzalcóatl, after a heavy rain storm in 2003, the ground washed away to reveal a vertical shaft almost five meters wide that travels 15 meters deep. After tons of dirt were removed, that entrance shaft led to a corridor as long as a football field. In the dirt and debris elaborate wood and jade masks were found, along with sculptures and mysterious, hard-clay spheres, termed 'Disco Balls' and covered in yellow pyrite, or fool's gold. Also, a pool of liquid mercury was discovered in the tunnel.

The pyramids of Teotihuacán increased our interest in Erich von Däniken's theories. More experiences during our sabbatical in Mexico would further our belief in Extra-Terrestrials (ETs).

THE BASILICA OF THE
VIRGIN OF GUADELUPE

On Sunday morning, we found throngs of church goers on their knees moving across the court-yard of the Basilica of Guadalupe. The hundreds of worshipers recited prayers as they slowly moved over the stone pavement. A combination of the immense weight of the cathedral and the soft earth of the historic, original lake bed had caused it to sink and tilt. We wandered the court yard and saw a dry fountain with crutches stacked beside it. It was a spring of healing water, now dry. Bill had two invalid sisters and we wanted to carry home a bottle of the water. As we meandered, we re-entered through the wrong door and encountered religious dignitaries in the sacristy. Although hardly dressed for the occasion, in shorts and T-shirts, we were invited to a personal tour by the cathedral's top religious men.

Basilica of Guadalupe

Our guide explained the cathedral had been built to honor the story of the Virgin of Guadalupe. Bill and I noticed there was a lot of magic everywhere in Mexico, not just deep in the mountains or jungles. More of the ancient alien/ET Vibe has an edge in Christianity.

The story of the Virgin of Guadalupe began before many Mayans converted to Christianity. The Virgin Mary appeared to a local Indian named Juan Diego in December of 1531. That's only a decade after Cortés conquered the Aztecs. Seems the farmer, who'd been an early convert to the Christian faith, heard beautiful sounds he equated to angels singing. A woman's voice called to him from the hill. The woman appeared to be a smiling teenage Indian, like Juan, emitting a warm, bright light that surrounded her. She was dressed in a peasant's red and gold embroidered dress and a turquoise shawl covered with radiant stars. The unique part was she floated above the ground on a disc held by another angel. That should have been Juan's main clue something different was happening.

In the language of the Indians, she explained she was the Mother of God and requested a church be built upon the spot. When that happened, she would remedy the local peoples' suffering. The cathedral was eventually built on Tepeyac Hill, about a half-hour drive in reasonable traffic from Mexico City. The hill was known as an ancient location for worshipping the Aztec Goddess, Tonantzin, a version of Mother Earth. All the forays to the bishop must have been long treks for Juan. The doubting bishop refused to believe the tale and wanted another sign before he would gather the indigenous slaves to begin construction of a church.

The Virgin made a second appearance and again Juan returned to the bishop to request a church be built. The bishop asked questions, yet still doubted the story of the poor farmer. He asked for a sign and instructed two priests to follow Juan. The farmer shook the men the bishop had sent and disappeared. Juan visited the Virgin and she ordered the farmer to return the next day. Juan's uncle was ill so he missed the meeting. The very next day, he encountered the Virgin again and she told him not to worry and asked him to pick flowers for her. He picked the flowers and she arranged them on his cloak. He carried them to the bishop.

Juan opened the cloth and presented the bishop with the flowers. Seems the bishop had prayed that roses be the sign and the flowers that fell to the floor were the fresh Castilian variety of roses. Every one present saw the image of the Virgin was imprinted onto the cloak. The cloak bore a spiritual message, easily understood by the natives. The local gods had failed to repel the Europeans and change was imminent. Eight million Indians soon converted.

The same day, while Juan was with the bishop, The Virgin appeared and healed his sick uncle. The Virgin told the uncle that she would like to be known as Santa María de Guadalupe.

We saw the tapestry hanging behind the altar. Although the cloak is of woven cactus fibers, which would usually rot within decades, it has endured five centuries. The kicker is, the cloth has been constantly studied. In 1938, Richard Kuhn, who won a Nobel Prize in chemistry, discovered the image isn't from natural colorings. In 1979, an infrared study showed it was not paint at all, but what was termed 'iridescence'. More ET vibes.

A Peruvian environmental scientist magnified the image's eyes 2,500 times. What was revealed was phenomenal. Within the Virgin's eye, is almost a Polaroid image of the moment the cloak was shown to the bishop. Juan, the bishop, and four others are in what appears to be the actual scene. In the pupil, is a second image, much smaller, of a scene of an Indian family.

Shroud of the Virgin of Guadalupe

Everywhere within the walls of the cathedral were plaques of devotees who had had their prayers answered. In 1921, a radical attempted to destroy the cloak with twenty-nine sticks of dynamite in a flower-pot placed directly below the sacred cloth. The blast destroyed a marble rail, twisted a metal crucifix and shattered windows throughout the old Basilica —but the Virgin's cloak was undamaged. Bill and I marveled at the twisted crucifix and still contemplate that mysterious wonder of Mexico, one of many.

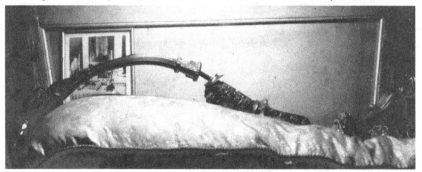

Crucifix bent in Basilica bombing

ROAD TRIP: HUAUTLA

BULLFIGHT

Our Sunday moved, as with many Mexicans, from church, to the amusement park, and then to the bull ring. Our Fodor Guide directed the VW bus to the Plaza de Toros. Few street signs and jammed traffic parked us about twenty blocks away. This is the world's largest bullring seating over 40,000.

It was definitely one of those days, as if Hemingway were lurking in the crowd. The matadors easily won the first two fights. The entire bull fight was obviously fixed, with the toreadors lancing the poor bull's neck to reduce the mobility of its lethal horns. But a slight rain intervened. The third matador slipped, as he planted his feet to sink his sword between the bull's shoulder blades and was gored in the stomach before being tossed over the horns. It was classic.

It must not have been a serious wound —since he returned. It was explained that if he could walk and did not confront the bull, his career would be finished. The toreadors did their best to immobilize the bull's head, but the animal bested the matador again, goring his thigh. Courage carried the fight to the third round and the bloody matador stood stalwart and slid the sword home. The bull crumpled at his feet and the crowd was ecstatic. Hats flew and beautiful women tossed roses.

Matador Down

MT. POPOCATÉPETL

After the pressure of a packed city, quieter surroundings with better air quality sounded like a pleasant change. Sam exited this adventure at the Mexico City Airport. He told us later his flight had been delayed a day and he'd slept on the floor. We had a route planned, but couldn't get adequate directions, so we drove in circles. After three hours, we finally stumbled on a route out of Mexico City by pure luck. The VW van rolled south towards the second highest mountain in Mexico. Mt. Popocatépetl (a mouthful pronounced pohpoh-kah-'te-PET-uhl) was a dormant volcano. The previous eruption had occurred in 1947. The crater was reported then to be filled with a frozen lake. Sounded safe.

This leg of the journey led us through Cholula the home of the largest pyramid in the world. This one is so big it dwarfs the huge church that crowns its top. This pyramid is not cleared of dirt like many others. Excavations have determined its dimensions making it the third largest, by volume, in the world. The prized guides told us that this pyramid had been built over other existing structures at least six times. The base of this pyramid is four times bigger than the base of the Great Pyramid at Giza. In the 1930s, explorations into the hillside found the pyramid structures and more tunnels. Lots of interesting relics and two unique paintings have been found in the tunnels. Quetzalcóatl is still remembered at the pyramid during both the spring and fall equinoxes.

Popocatépetl (also known as Popo) lies south of Mexico City. This snowcapped peak is joined by Mt. Iztaccihuatl, 'the Sleeping Woman' (pronounced e-stak-see-WAH-til). There's a fable that explains why the mountains are joined. A tribe was tired of paying tribute to the Aztecs and decided to revolt. Iztaccihuatl was the beautiful princess daughter of the chief who was in love with the handsome warrior, Popocatépetl. The chief promised a wedding when the warrior returned from battling with the Aztecs. While gone, a rival told the princess that Popo had been killed in battle. The princess succumbed to sadness and also died. Popo returned in valor only to learn of her death. Ten hills were amassed to make a notable grave that no one would ever forget. Popo now kneels with the smoking torch of the volcano, watching her sleep.

Our trip south was well timed. Popo heated up and began erupting in 1994 and hasn't stopped. In 2016, the volcano had lava flows.

The snowcapped, majestic mountain came into view about an hour from the city, but it was a difficult drive to get closer. At 5,400 meters, it stands out. In those days, no buses made the journey. The only option was a small train that came reasonably close, stopping at the tiny village of Amecameca, five miles from the base. After fast-paced Mexico City, this village was the desired sleepy and picturesque respite. At the base of the mountain, it seemed like a Mexican version of the Swiss countryside. The neat, brightly-painted buildings, the colorful train, and particularly, the clean, cool air were refreshing.

Lunch was at a sweet little restaurant. The special of the day was bistek, which I translated as beef. The steak was an inch thick and covered the plate. Bill winked at the waitress and she smiled as we sliced off a chunk and began to chew, and chew, and chew. The older woman watched, as it was hard to break the meat down enough to swallow. We weren't strangers to stringy, tough meat, but this was different. I pointed to the plate and asked "Bistek?" The older woman smiled and shook her head and answered, "No, caballo." We were having our first-and-only meal of horse. Somehow, our fear, and being timid at eating what was probably an old or sick horse amused the giggling, snickering ladies. Beans and rice again filled our bellies.

Bill at the Base of Mt. Popo

We were off to discover Mt. Popo. In that territory, the VW was a luxury machine, even if it never got out of second gear. The entrance to the national park slowly wound around the mountain through a forever pine forest. Everything was green with moss. The sides dropped off so quickly that parallel to the road were the tops of immense trees. We arrived after the six pm closing. Our bus parked along the road near the entrance and spent a chilly night.

Our energies were recharged by the morning vista. It was time to give the van a bit of reorganization. Everything was off-loaded and repacked as I changed the plugs, the oil, and cleaned the filters. The park rangers engaged us with info that we should beware of snakes and scorpions, and that there were nearby trout streams. The snowcapped expanse held us in awe until we gained confidence by talking with a young lady whose friends were already climbing. Not prepared for snow, we drove down to Amecameca again to find warmer clothes. No gloves were available.

The next morning, we rose early and started off to the trail head at the tree line. We rented ice poles that doubled as hiking sticks and metal crampons for the ice we were told we'd encounter. The trail was set up in three parts with small severe weather cabins marking the stages. At the second cabin we found the snow and started up at what seemed to be a 45-degree slope. By ten in the morning, we were surrounded by a cloud with no visibility to advance or retreat. We sat and shivered for about an hour as the fog became a snow storm. The scene cleared when the sun warmed enough to elevate the cloud.

The view of the frozen crater lake, reflecting a myriad of blues, was worth the effort. We toasted with a flask of tequila, feeling like we'd conquered Everest. Our triumph deflated when a group of school kids with their teacher ran up the crest in their gym shorts.

With the thrill of adventure pumping, we plotted a route to follow the trail Cortés' took after his butchery of Cholula. Cortés gathered all the Amerindian royalty in the center of the village and then attacked the weaponless villagers. Helped by another tribe, almost every inhabitant of Cholula was killed. This may have been a successful and frightening message to the Aztecs. I doubt if he took time to make the extra climb to view Popo, but the path he marched to attack Mexico City was our route of descent between the two mountains. The road was slightly more than a wash out, but it looped through another moist, green forest. It probably hasn't been altered much since the soldiers' march from Puebla to the Aztecs in Tenochtitlan in 1518.

Our map showed a dirt track, named Cortés' Trail, which crossed between the two mountains. The views, as we dropped elevations, were truly gorgeous. The huge, tall trees could have been virgin forest. They provided enough shade to keep the grass short, so it looked as if the woods had been manicured. In an American moment, it was Kodak time. Bill dismounted, almost plodding onto a cow pie. Growing out of the pie was a clump of parasol mushrooms, which could have meant psilocybin. They were tiny, with slender stalks and blackish caps that measured two to three inches. Our parking area was a 'shroom site. We had no idea how to discern if they were fully grown, ripe, or how many would be a safe dosage. A few of the fungi were sampled. The VW beamed the remaining trip down the mountain.

Monique's friend had told us to eat two full banana leaves. The next question was, how many 'shrooms fill a banana leaf? Was Ernesto correct about the pastures and the cow pies? Had it been a prank description? Were they even edible? Our curiosity overcame our reservations and ignorance.

The VW was the only vehicle on the road. Occasionally, we would stop for a small herd of cattle or a couple of horsemen. Bill drove on while

ROAD TRIP: HUAUTLA

Earl and I sampled our find. The mushroom wasn't the usual tasteless, but acidic, slightly bitter. We parked and built a campfire next to the road before dividing the remainder. By the time the sun was low, we were high. Some local cowboys rode up and helped drink our tequila. Bill thought he saw Cortés' bronze conquistadors sitting by the fire —and that's about as far out as we got. The fungi weren't as potent as expected, but they did wet our appetite for future experiments.

RESERVOIR CAMPING

Not yet dead, or noticeably dying from mushroom poisoning, in the morning we rekindled the campfire. Strong coffee, made with two heaping spoons of instant coffee, helped clear the hongos' cobwebs. After pondering over the road map and Fodors', and after back tracking to Cholula, we could see there was a lake. On the VW's roof rack was a six-man inflatable raft we used for white-water rafting in the mountains of Pennsylvania. We cruised through Cholula and Puebla. The first lake we encountered was only a marsh. At Temescal, we found a reservoir that fuels an electric generator. Our van and raft became the biggest things that probably ever happened at the shoreline village of only thatched huts.

Raft at Temescal Reservoir

Our south-of-the-border vacation adventure had been fairly hectic. It was decided that we'd row out to the island in the center of the lake and camp for two nights. It looked so tranquil and perfect for lounging. Once the hammock tents were strung, we started gathering firewood. There were no 'shrooms growing on the small island. Flipping over a stump, we found the biggest snake any of us had ever seen. As the sun fell, we made certain the fire blazed. Strange noises, including the crash of Bill's hammock, marked a restless night.

A dark sky threatened the following day. Once we were somewhat secure in the hammock tents, the storm descended on us. The nylon roof kept us dry. It sagged as it filled until the cold rain-water touched our bellies. This was a bad night and the overturned raft proved to be the best shelter. The wet ground ants found our tender flesh. It wasn't pretty. After a torturous four hours, we cashed in and rowed across the dark lake to find a dry bed in the van.

We woke up to jabbering faces peering in the van's windows. We'd forgotten a few items on the island in our haste and rowed back. The Federales were waiting at the van on our return. The local fishermen believed we were taking their fish.

Hammock Tents at Temescal Reservoir

VERACRUZ

Leaving Mexico City, we'd decided to head to the Yucatán rather than to Oaxaca. The reported land of magic mushrooms would probably signal the ending of our trek. On this section of our travels, we saw our first auto accident. It was horrific. A Ford Mustang topped a hill and smacked a loaded semi so hard that it knocked the front wheels off the truck.

Veracruz seemed like a reasonable stop on our drive to the Yucatán. This was the most attractive town we'd visited. Every building was white and spotless, but not as inexpensive as the rest of Mexico. Veracruz was upscale and many of the old stucco houses were topped with unique domes. Any moment we expected to see John Wayne or Robert Mitchum strolling down the street. Veracruz was sparkling clean and far removed from the population-pressure we'd experienced in the capitol city.

Of course, we found a locals' bar and a few other gringos. We proceeded to drink too much tequila—creating expectations of Burt Lancaster and Gary Cooper swaggering through the saloon's swinging doors. I'd seen the old motion picture with Lancaster and surprisingly, the city's skyline hadn't changed much since the '54 flick, especially the bar's ambiance.

The nice evening invited a stroll to find a place to relieve ourselves. Huge Castillo San Juan de Ulua became our goal. It had been an early Spanish fortress built in 1565. Captured during in the Spanish American War, by then it had become it one of the world's filthiest prisons.

Toking what remained of our smoke, Bill and I conversed while we rambled. Walking and talking, we paid only enough attention to our surroundings not to misplace the VW. The beers had darkened the street and nature strongly called. Traipsing from dimly-lit to dark, a suitable tree was found for bladder relief. Then our stroll took us along a well-lit, modern concrete wharf where the boats were replaced with ships. At the end of the dock, we had to turn right, and never noticed we'd passed through an open gate. On all sides were large weapons, cannons, and machines guns dismantled and lying around. We climbed the stairs and sat in the gun port. We waltzed around as though we owned the place and stumbled onto a circular staircase that led to a high, corner tower. We weren't trying to be quiet.

As we scanned the inside of the old fort, it was as if we'd wandered into the Mexican Army's junk yard. We climbed down and decided to cross the eerie courtyard. The highest tower beckoned to be climbed. It was then we realized these towers held real guns, defending the harbor and naval installation. As we passed a couple lighted doorways, we heard "Hola".

The voices were barefoot in boxer shorts and ragged tank tops, but each had an old, bolt-action rifle aimed at our bellies. The men were guards who must have been snoozing until they'd heard our voices getting louder with the sight of all the artillery. Again, our limited Spanish was a huge problem, but their meaning was clear. So were our apologies for whatever we had done. We were somewhere we shouldn't have been. They'd probably smelled our smoke, yet that didn't seem to be a problem. They didn't appear to be Federales. With hands up, we explained our stupidity and cautiously retraced our steps.

Minutes of jabbering and gesturing at the many big ships that we hadn't observed in the darkness, we finally understood we'd staggered into a Mexican Navy base. They shouldered their rifles and politely showed us the way out. We found another bar for a suitable night cap and laughed at our luck. Mexico was not dangerous in 1974, but we'd been stupid. Lesson learned again for the umpteenth time: tequila, beer, and weed only in moderation.

Earl decided to head home from Veracruz. When we'd chosen to take the eastern route to the Yucatán rather than drive directly to Huautla with the mushroom cult, he wasn't pleased. He'd met two girls and that was the last we saw of him. The sad truth was, we didn't know exactly where Huautla was located. Veracruz was at the junction of possible routes, mysterious Huautla or the Mayan Yucatán. Beaches and pyramids among the ruins won. After the hectic touring of a few of the maximum sights in the days before Sam departed, fighting packed highways, it was time for some scenic ruins. Bill was definitely not for pushing into another set of mountains.

With more sleeping room in the van, Bill and I headed along the coast, east to stunningly beautiful, fresh water Lake Catemaco. We met Luis, an urchin guide, who took us to the mineral springs at the village of Coyame about ten kilometers away. When we were staying at a hotel in Mexico City, Coyame was the bottled water. Drinking regular water out of a faucet was a definite no-no, so the first two bottles were free on the hotel's night stand. It was natural sparkling water.

Luis directed our van to a large wooden building. The locals energetically welcomed us as probably the first tourists who had wandered that far. Cases and bottles were everywhere, but there wasn't any visible industrialized bottling equipment. It was all pretty plain stuff. Two men took us into the basement and we saw just how basic it was. They opened a trap door in the floor and there was the water. Their job was just to fill bottles from a hose and cap them. The natural mineral gas rising off the water was intoxicating. We filled everything we had with free mineral water, about fifteen gallons.

Our guide wasn't finished yet. Further down the road was the Eyipantla Waterfalls. Several streams joined and dropped a hundred thundering feet. The boy led us up the carved steps—and after we had worked up another tropical sweat, we cooled off in the pool, catching the spray from the falls.

Eyipantla Waterfalls

We didn't know it then, but the Catemaco area of Mexico has been well known for witchcraft since before the European invasion. It seems to be a center for extreme Santería worship with both good (white) and evil (black) brujos. Santería is a combination of native rituals and beliefs combined with Catholic rituals. A bit further west in Santiago Tuxtla, where there is a huge Olmec carved stone head, is what's also known as a center for healing. Every part of Mexico seems invested in magical lore, ancient aliens, or present day UFOs.

With fifteen fresh pesos in his pocket, we parted ways with our young guide. Luis had done an excellent job. Our next stop was the village of Avayucan for lunch and then on to Villahermosa. The summer days were long and the roads were good—with little traffic. The thick bush crawled to the edges of the road and our imaginations roared with what ruins might still lay hidden throughout our three-hour trek. Our bus had no tunes, only endless chat. We were on the trail of the Mexican adventure that had been our magnet.

YUCATÁN

From Villahermosa, we turned our route north and followed the gorgeous coastline through Tabasco and into Campeche. We hopped on to ferry boats at Frontera and San Pedro and crossed over onto Isla Del Carmen. The main town wasn't very big, but the main industry was building beautiful wooden fishing boats. At that time, the cost was only a few hundred US a foot. One fisherman decided to be our 'friend' and let us know weed was available. Fishermen, he reported, always had weed. The government permitted them to smoke to keep them happy and the fish coming. He didn't have any, but if we gave him some money he could get us some. We didn't, so he didn't.

No other cars on the roads made this a private island for exploring in our VW. We found a nice stretch of beach and made camp only to be attacked by a zillion fleas. After a thorough salt water bath, we fled to Champotón. Back then, it was a small, mellow fishing village. Daily doses of saltwater swims energized our trek, and usually a few speared or netted fish became part of our dinner. The beaches were alive with crabs, iguanas, and snails. At night, we would hunt crabs that wandered onto the road. We'd chase them, bop them with a stick, and then throw them into the boiling pot. We ate well.

Tarantula

The following day, the weather darkened as we drove to Campeche. After inspection of the archaeological museum, we roared on towards Mérida. We stocked up on fruit at the market. I found a big, sweet papaya and added that to our usual lunch. The map showed a more interior road that passed through the Mayan site of Uxmal. Route 180 had been good to us, with plenty of beaches since Villahermosa, but the Mayans were our new target. There wasn't much to see except that this was real jungle—or, as real as it gets for a pair of guys from Pittsburgh. That night, black spots appeared on the road: tarantulas. These frightening spiders were as big as our feet. At a sign for a cave named Grutas De Xtacumbilxunaan (try to pronounce that!), we parked, sealed the van, and crashed for the night.

The next morning, we encountered a few people with broad foreheads, high cheekbones, and prominent noses who looked to be descendants of the Mayans. They were staring into the van's windows. Our night cap of tequila mixed with Kahlúa, mineral water, and lime juice had induced a deep sleep. To our uninvited visitors, we were a shocking vision. We were on display. We hadn't really paid attention to where we'd parked, but had just had taken it for granted from the official-looking sign that it was some type of government-maintained location. The dense jungle opened to a clearing and we'd missed the two traditional round-houses. These homes were unique oblong structures with thatched roofs. This style probably hasn't changed in design much over the centuries.

Yucatán House

Aside from the excitement of our unannounced, early-morning visitors, my legs had sprouted red blotches. Something was causing an allergic reaction, but I didn't think of that until I'd eaten the remaining papaya for an easy breakfast. By the time we hit the Mayan site at Kabah, the blotches had formed into big water blisters. I'd been so dumb; the fruit had caused the blisters. That was the only papaya ever in my life to cause a reaction. When we hit the Mayan site of Uxmal, my blisters made it painful to climb the tall pyramids. We'd taken this road specifically to see Uxmal, and the place was beautiful. Blisters or not, we made it to the top. It had been very good luck to meet the Penn State girl at Tula. We made it a rule to have an official guide at every big ruin. Each official conducted tours walking slowly while spouting a lot of facts.

Uxmal translates to 'built over three times'. Several of the pyramids are the compilation of many previous pyramids. It seems most of the pyramids in Mexico are built over, like the biggest one at Cholula. At Uxmal the first big one is the Pyramid of the Magician. It's almost a hundred and twenty feet tall, so he must have been quite the magician. It does not have a square, but instead a unique elliptical base with smooth, rounded corners looking like a weird, inverted funnel. Our guide reported this pyramid was built on top five previous temples.

Pyramid of the Magician

The name comes from the legend that the king of Uxmal knew he would be replaced when he heard a gong sound. This was centuries before Chuck Barris' The Gong Show. A dwarf, who was also a magician, came to Uxmal after the gong sounded. The king, like most kings, was not ready to relinquish power and ordered that the dwarf be sacrificed. The citizens of Uxmal thought the dwarf might be able to offer a better future for their city. The king decided to issue a series of unreasonable tasks. If the tasks were completed, the dwarf could continue breathing. One seemingly impossible assignment was building a unique pyramid. The dwarf completed it overnight. Then the dwarf and his mother went to war and conquered the Mayan king. The dwarf ruled the city and prosperity followed.

This dwarf saga could be why two other translations of Uxmal are 'the city of the future' or 'city of promise'. It could also be the origin of the magic that exists in Mexico. Again, the possibility of ancient aliens arises, since the west stairway faces the summer equinox. The dwarf may have been a short ET since the site has excellent geometry. Uxmal has the best, most extravagant and most detailed carvings we'd ever viewed. This site is very extensive due to the underground water supply that linked it with Kabah. Supposedly, 'the Uxmal builders' aligned with Chichén–Itza's builders and together ruled the Yucatán peninsula until they succumbed to the onslaught of the warrior Toltecs.

By late in the day, after we'd toured the Governor's Palace, The Nunnery, and The Palace of the Doves, my papaya blisters had disappeared, but my butt was chafed. Driving east, the scenery changed from a humid jungle to a scrub-wood wasteland. As we drove to Mérida, we occasionally saw a tall rock outcrop and wondered if it could be an undiscovered Mayan site. There were no rivers, which is why the underground streams were so important to the rise of the Mayan culture. Both sides of the highway had impenetrable scrub brush with absolutely no side roads.

Our days were packed with people, places, and historic sights. Mérida was a beautiful city, reminiscent of what Mexico must have been fifty or more years previous. Horses were hitched to posts outside stores. The outlying streets were dirt. We found a wood shop cutting boards from trees and bought some very dark, heart wood not unlike ebony. The center of the city was an aged square, or zocallo, surrounded by restaurants packed with French tourists. The market was mainly peasant clothes and some nice leather goods.

Chichén Itza was farther than expected, about a hundred miles west of Mérida. Its name translates as 'edge of the well'. This well was the open cenote that held our interest. We had read stories of divers who'd gone down into the black water to find valuable artifacts the priests had tossed in along with human sacrifices. It was more than we had anticipated and worth every minute and expense of the trip. This Mayan site had all the fabled stuff: a ball court where losers had their heads chopped off, stone walls carved with skulls and snakes, a deep steep-sided well where they tossed sacrifices, another building where jaguars ate hearts, and a stone ET domed observatory.

Chichén Itza scene

Sacred Cenote

ROAD TRIP: HUAUTLA

There was the Great Ball Court where men played for their lives. Outside the ball court was the Skull Wall—perhaps where the losing players' decapitated heads were displayed. I was glad we weren't high since some of the carvings were grim. One panel showed a winning player cutting off the head of the loser with a flint knife. From the loser's neck, blood spurted in streams that transformed into six snakes. The seventh blood squirt was a vine symbolizing the roots of the sacred tree at the center of the earth. The Temple of the Jaguars is connected to the Wall of Skulls. It is suspected, their big pet cats fed on the bodies of the victims. The rings of the ball court had been built from stones carved to resemble the winged serpent chasing its tail. The winged serpent seemed to be represented everywhere.

Another pyramid was dedicated to the Mayan feathered serpent, Kukulkan, a version of the Toltec's Quetzalcóatl. This is the type of pyramid one imagines when thinking of pre-Columbian America, with stairways on each side. Each stairway has ninety-one steps, and counting the top platform, the steps total the 365 days of the year. It is a combination of several pyramids built atop each other. The final pyramid reaches one hundred feet with nine terraces. As we climbed in the humid heat, the terrace rest stops were necessary.

Chichén Itza—Temple of Warriors

Almost at the top of the pyramid was a cut out where an exploratory excavation had located the most recent underlying temple. Inside one of the two unearthed rooms was a red Jaguar, inset with big, jade knobs representing the cat's spots. This sculpture that is purported to be a throne—uncomfortable at best. In the other room was a chacmool. The jaguar looked more like the chacmool we'd seen in Tula. There was a good reason for the similarity, as one of the guides stated that Chichén Itza had become great only after they either joined forces with the Toltecs or had been conquered by them. The Temple of the Warriors looked like a copy of Tula's Atlantes and the column statues had the same name. It consisted of a large, stepped pyramid fronted and flanked by rows of carved columns depicting warriors. The feathered serpents' heads, at the base of one of the north stairways, bore a close resemblance to what we'd seen in the pyramids of Mexico City. Another potential ancient ET engineering experience—the builders had been able to design and orient this pyramid so that on the vernal and autumnal equinoxes it would cast a shadow onto the adjoining balustrade, which appears to be a serpent twisting over the columns. That is very well planned for a Stone Age culture.

The temples at Chichén Itza are unique in their shapes and provide serious adventure-imagination stuff. Uxmal has the pyramid with the rounded edges, but the Observatory and Sacred Cenote in this site puts it over the top. Add the jungle surroundings, and all you need is Indiana Jones. El Caracol is the official name of the observatory. The round stone dome is incredible—hardly built by a primitive culture. In various parts of Mexico there are significant intricate structures that would be difficult to build today without huge equipment directed by precise measuring tools, both developed only during recent times. It is not difficult to believe that the winged serpents could have been the ancient aliens in flying crafts. The lack of proper terminology and a total lack of comprehension of air travel could have led to the winged serpent descriptions. After all, the ancient world had a lot of fables and legends concerning flying snakes and their cousins, the dragons.

El Caracol's dome was definitely constructed to view various parts of the night sky. It looks like it was built for a telescope, and has some pretty amazing, advanced architecture. The very top of the dome is flat so someone could potentially sit above the trees and watch the night sky, or see other tribes about to invade. The northeast and southwest corners align with the sunrise of the summer solstice and the sunset of the winter solstice.

Like many other ancient civilizations, Mayan observers followed the planets across the sky. A spiral staircase climbs to the top section of the dome. That's why it is called El Caracol, 'the curled snail'. The remaining windows are aligned with the movement of the sun, moon, and some of the brighter stars.

Archaeologists and astronomers believe this observatory was constructed to primarily follow Venus, known by the Mayans as 'Chak Ek'. This easily seen planet was significant to Mayan priests who considered it the twin brother of the sun. The Mayans supposedly were convinced that Venus was the heavenly residence of their main god Kukulkan, (recognized by other

Mexican pre-Columbian cultures as Quetzalcoatl). El Caracol was probably where priests watched the planet's transit, moving between the sun and earth, and computed the path of Venus.

Mayan Observatory

An American was its first owner and did a lot of the exploring, according to our guide. Edward Thompson was the US Consul to the Yucatán. This guy found all sorts of great stuff. His explorations started in 1894 and continued for thirty years. He was the first to search and dredge the Sacred Well. In tombs, he found the first examples of Mayan cloth and parchments.

The Sacred Sacrificial Well, or 'cenote', is a perfect setting for high adventure and exploration—a fantastic reality of the Ancient Mesoamerican world. It is a large, natural cylinder that's two hundred feet in diameter. It's almost perfectly round, with sheer walls ninety feet above the blackish water. Just as within the Mayan sites of Kabah and Uxmal, subterranean water was the key to their development. An estimated seven thousand 'cenotes'— or natural sink holes—have been discovered in the Yucatán's limestone. Over the years, divers have found all sorts of objects in this Sacred Cenote that are believed to have been offerings to the gods, including gold, jade, and human bones among them. Were human sacrifices tossed in alive? Did people watch them struggle with the impossible climb and then drown from exhaustion?

The Mayans had also built a top-shelf steam bath that worked by heated rocks. Perhaps it was used to rinse off the blood of their sacrifices. There is also a place to wait until there was suitable space inside the steam room.

Either before or after, there was a fresh water bath. These guys didn't live too rough; they had an abundance of fresh water and enjoyed it.

One more ET aspect was pointed out by our guide at the Platform of Venus. Unique to Chichén Itza, is that most of the buildings conceal secret rooms considered to be tombs. The second explorers, in the 1880s, were the Le Plongeons, husband and wife. They found urns with cremation ashes, and a hidden room below the floor of the Venus platform that had yet another Chacmool. Almost two hundred large stone cones were discovered beneath the Chacmool and no meaning has yet to be attributed. I'm surprised no movies have been made about Augustus and Alice Le Plongeon, as traveling to these sites in the 1800s must have been incredibly rugged. Their early photos, which were displayed at the site, were astonishing. The photographs depicted the colors that the Mayans painted everything, mostly reds and blues.

As with most cities, beside the sparkling new buildings is an older, more run-down section. The pressure from clamoring tourists pushed us down a jungle trail to the 'old city' of still unrestored ruins. We were damn surprised when our trail became small gauge rail tracks. Seems the ruins had been accessible by a steam locomotive from the 1880s until the 1960s. That would have been a wild way to travel through the Yucatán. We were disappointed when the old city was just piles of rubble. If no one had been around, Bill and I would have been moving rocks, looking for artifacts.

Instead we decided to say goodbye and retrace some steps to get back on track for Huautla and the magic mushrooms. Two Americans hitched a ride as far as we were traveling that day. They were returning from the south coast of the peninsula. They'd been camping at several fresh water lakes and the famous seven shades of blue lagoon at Bacalar. It is a lake/lagoon formed by seven cenotes and the color of the water varies at each. The hitchhiker's smoke was so good it made returning to Mérida in the sunset rain storm hazardous, but the camaraderie helped make it enjoyable. To finish a near-perfect day, we gorged on turkey in black sauce that was hot as Hell! Many cervezas were the best relief.

With Oaxaca now in our sights, we tried to bypass Campeche by visiting the Mayan site at Edzna, with the five-story pyramid. The road was too bad and again we had to backtrack through to Champotón. We decided our next destination was yet another Mayan site at Palenque. We'd seen the north coast of the Yucatán and more trucking on the interior roads fed our lust for adventure. Most of our travels took at least two days. We'd been eating a lot of rice and canned beans cooked over campfires and drinking lukewarm cervezas. Farts were our sound track.

Palenque appeared with the stars of the night sky. Now in the state of Chiapas, we did the usual—sleeping in the parking lot and awakening to the now usual, smiling faces of locals. We wondered what they thought. Was our blue and white VW transporter so unique, or was it Bill's loud snoring that was the attraction?

After a breakfast of last-night's rice mixed with fried eggs and super strong instant coffee, we attacked the ruins. Palenque was very different from the other Mayan sites we'd already visited. The Mayans were here early, around 250 BCE. That predated most of the other sites by five hundred years. Its location may have protected it, as it is believed to have existed longer—almost a millennium!

Palenque Ruins

Palenque appeared smaller than the other sites because very little—maybe only five percent—of the site had been cleared. Still what was there was exquisite. Amid real jungle, Palenque looked like a tropical paradise. The inscriptions showed handsome people and plenty of clear lettering that had yet to be translated. The four-story observation tower was impressive, as was the Temple of Inscriptions. We'd come to see Erich von Däniken's famous tomb of the Mayan astronaut. Several tombs had been uncovered, including one with a jade mask.

The lights weren't working that day at the Pyramid of Inscriptions as we climbed down the stone steps to the tomb room. Always prepared, we had our trusty flashlights. As luck would have it, French tourists had a camera flash and I got a few shots of the famous tomb slab cover. It has now been translated and is known as Pakal's Tomb. Pakal is interpreted to have ruled the longest of any chief anywhere in the Americas: sixty-eight years. That's a lot of sovereignty for any time—making him thirtieth longest ruler in world history. Plus, a ripe old age was only forty in those days. He was buried in a stone sarcophagus similar in style to the Egyptians.

Palenque Wall Carving

Again, where was Indiana Jones? This rare Mayan tomb had a hidden doorway in the floor, sealed by a huge flat stone. It was revealed by Mexican explorer Alberto Lhuillier in 1952 after four years of removing rubble. The skeleton of Pakal was still in the stone tomb wearing a jade mask.

The carved tomb lid shows Pakal seated—and according to ancient alien believers, at the controls of a space ship with breathing gear hooked to his nose. Depending on who is translating the lid's inscription, it supposedly describes how his soul would return to the stars by riding on the 'Monster of the Sun'. Everything the ancients did, they combined with astronomy. What we'd seen in all the sites involved a combination of the efforts of highly skilled mathematicians, architects, engineers, and astronomers. Add to that a load of laborers, tons of food, and some very strong tools for intricate work. Why couldn't they—and all the other pre-Columbian cultures—have been clued in by ET's?

Understandably, all the Mesoamerican cities revolve around a water source. Palenque has a beautiful stream running through it. With the stream and the surrounding jungle, this place had an acoustical hum. At least I heard it. It was like the sound you make when you meditate: 'oomm'. Palenque is an extremely different place. It's easy to find a peaceful feeling there.

Leaving Palenque with more questions than answers, we helped a California couple get to Acayucan as they were heading for Veracruz. Our goal was to get some answers on the cult of the magic mushroom. We still weren't certain of much—except the town was located in Oaxaca. Our map showed another sound-a-like town, Cuautla. Confusion reigned.

Palenque Pakal's Tomb Cover

The following day was a Sunday and we were pushing to get into Oaxaca. A day of rest in Mexico means a day of drinking. Too many Mexicans drive drunk. Heading to Tuxtepec, at the state line we saw many drivers weaving over the roads. One semi had launched off the road and down a hillside, and another had jackknifed, blocking traffic.

The wait for a ferry across the Papaloapan River was an hour. Our van was moving slowly, but a walking farmer moved much slower. We gave him a ride and hoped to gain some insight on the local area. He didn't know anything about mushrooms or Huautla. The road to Tuxtepec followed the winding river. Our passenger insisted on buying us a drink at his hometown bar. All his farmer buddies were smashed and wanted to shake the gringos' hands. We tipped a few beers before we found a camp site next to the river.

OAXACA

We'd hoped crossing the border, into the Mexican state of Oaxaca, would incite a fresh state of mind. We did the usual clean-up-the-VW-bus routine: give it a good wash and sweep. We tried to air it out every day as it was getting a bit too homey. A couple of guys sleeping on the cushions for weeks had rendered it just north of 'eau de locker room. Opening the sunroof every afternoon, we were optimistic that Mr. Hot Tropical Sun would burn off any unwanted bacteria and return that 'sunshiny fresh scent'.

The border was so elaborate that it had a shower facility unconvincingly labeled 'Employees Only'. While Bill was scrubbing, one of the border guards sauntered over and inquired if I would like to buy some marijuana. Obviously, I declined. The guard shrugged off my dismissal. As he left, he muttered that if we were searching for pot, we could find the absolute best quality in the state's capital, which was officially Oaxaca de Juárez, but almost everyone referred to it as Oaxaca-Oaxaca. The state border patrols were the best public relations personnel in Mexico. They were the cause of tons of greenbacks flowing into Mexico by way of black market commerce, bail, fines, and/or bribes.

Oaxaca Road Into The Mountains

ROAD TRIP: HUAUTLA

Our route climbed into foggy mountains. Every summit was draped in an eerie mist. Waterfalls seemed spaced a few hundred meters apart.

All of our surroundings were luscious, wet, and dark green. Everything was picture book fantastic, except for the stray cattle that kept blocking the road. While we waited for a small herd to pass, we paused for much-needed bladder relief. Cattle paths bordered the narrow road. Little mushrooms sprouted from almost every pile of cowshit. We, of course, gobbled a few and drove on. Every time we stopped, we found more 'shrooms. We always caught a buzz, but never really turned the bend to getting turned on tripping.

Gobbling supposed psychedelic mushrooms along a mysterious Mexican mountain road may seem like pure lunacy. It's difficult to translate the exact philosophy that was our operations manual. At every turn, another incredible view was presented, not very different from a Walt Disney travelogue. The mountain pastures projected a timeless serenity. Again, hard to translate—it was as though these visages were alive and whispering. Perhaps the mushrooms were the only Oaxacan conventional means of understanding the messages.

Oaxaca, the capital of Oaxaca, wasn't imposing. The city had unimpressive, two-story buildings built around a zocallo, or central square. Adequately parked, it was cerveza and taco time. At our bar of choice, we teamed up with two other America wanderers. Young Sally was a curly haired, 18-year-old Bostonian. Her mother's graduation present had been a trip to Mexico. We never discovered if her mama knew Sal was traveling solo. This was her coming-of-age passage before beginning college and entering the work-a-day world. From her tales of Mexican buses and trains, Sally had been extremely well educated in how the world turns south of the border.

Colorado Mike said he was in Oaxaca to buy blankets for resale in Boulder. He never swayed from the story, but it certainly seemed his intentions were more towards dried vegetable imports. A third young damsel, Francois, was French. As we filled the three in on our mushroom adventure, we opened our tour bus for those who would accompany. The French girl had better sense, or previous commitments. She was heading back east for her return to France. I've never understood these young women traveling alone in the Mexican wilderness.

Our invite to tag along was followed by their invite to share their hotel room. It was a long night on a hard floor. Early the next morning, the city market was perhaps the best I'd encountered. By noon, our van, with Colorado Mike and Boston Sally in tow, headed almost directly north on a small road. Fingers were crossed that our directions were correct.

ON THE ROAD TO HUAUTLA

According to our official Oaxaca state map, Huautla was only eighty miles north of the capital. It seemed it should be easy to locate, as there was only one road that led there. The map did not depict the road as an expressway. Once our van rolled onto it, we discovered another version of tourista optimism: never trust a Mexican roadmap. When the map depicts a roadway as a dotted line, it should mean an unimproved road. In our map's case, it meant a non-existent road. The first ten miles were a gloriously paved highway. After a determined right turn, our route became a typical Mexican composition of rocks and dust, with more than enough pot holes to traverse. We slowed progress to 15 miles per hour to avoid certain disaster. With the weight of four now aboard, the VW enjoyed crawling through the mountains.

In the arid, high-desert heat, we had the sunroof open. Our pace was slow and made minimal dust. Sally and Mike had perched on the roof luggage rack. The desert trails all looked alike. On the dashboard was our compass, but the map proved unreliable. At an intersection, it was difficult to determine which road was the main artery, as all dirt tracks were seldom traversed, except by horses. At every cluster of houses, we would inquire that village's name, but never once was one on our map.

High Plateau Desert

Sally was our lifesaver. A Mexican male will tell a beautiful, young woman anything just to get a chance to pitch his rap. She became our resident interpreter as we bounced from one cluster of mud homes to the next.

One wrong turn led us to an orange adobe church that trusted its parish enough to chain and padlock its doors, which were marked 'Welcome to all'. The women of this dusty village rushed our van to hear some news. Sally alluded to our adventures while trying to ascertain where we were and get directions to our mushroom destination. Unknowingly, I repulsed the women by taking off my shirt. It was impossible to apologize for a blasting sun that created almost unbearably dry heat.

One of the older women told Sally that Huautla was further north on the road. I'd learned it is difficult to accept directions from women who only left their villages in a predestined bus. We were stuck in the dust with no radio reception in the days long before iPods. Peanut butter and jelly on warm tortillas sufficed for our big-time lunch. Before the sun sank, all our get-high was gone.

Our official map of Mexico had two Huautlas in Oaxaca. One had the surname of Huautla de Jiménez. It had never occurred to Bill or me to ask the Freak Brothers with Monique to show us on a map. We were headed towards the one name Huautla, but when we got to where it was supposed to be, it wasn't. None of that village's people realized that Huautla, with no extra name, was supposed to be their immediate neighbor.

While I filled the bus with gas, Sally got valuable info from an older, seemingly-important local. His directions were, head into the mountains. Not to be led further astray, I walked to the market and inquired. A chiquita with a gold tooth blushed and asked if I wanted the Huautla with the mushrooms. She then assured me it was the 'de Jiménez' village.

Meanwhile, the older, more distinguished man had won the graces of my crew and he just happened to be going our way. We drove him out of town to his tone of lefts and rights. Suddenly, he commanded me to stop for two girls sitting on the curb. He hustled out and lifted one girl into the van, as she was paralyzed. After the other girl got in, we drove away. In the rear view mirror I noticed an ancient, weathered woman in a black dress chasing us shaking her fist. It dawned on me the old gent might be kidnapping her daughters to satisfy a fantasy. He hopped out at the town limits with no explanation except that the girls were going farther along our route. We deposited the two beauties at the next set of mud huts a little way up the road. I've wondered what was up with the woman in black. Perhaps she was an evil bruja holding the girls hostage. All the old man had told me was, "Never mind, drive on."

In the days before GPS, our expected one-day jaunt stretched into a second day as we realized that the eighty odd miles our map displayed had doubled. We began seeking a campsite while it was still daylight. Our one-van expedition pulled into the small village of San Juan de Los Cues. This was the largest village we'd hit since we left the capital. Our new navigator, Sally, had the best command of Spanish. As she asked the village's name, Colorado Mike hastily accepted an invitation from a local family for a cold beer.

Grins and sly glances began once Huautla was mentioned. It seemed everyone knew of the mountain town's specialty: hongos. The family's head intimated that he had visited Huautla only once in his youth. His trek had been part of a macho adventure and he had consumed the mushrooms with a sacred woman. His words of wisdom were not to consume the mushrooms with any other drugs, including coffee, alcohol, or tobacco. We should not take the mushroom experience casually either —as we could easily end up insane. He agreed that although everything we'd been told sounded crazy, it was all true.

That afternoon, we were definitely drinking. Sweet-faced, blond Colorado Mike became the comedian and appeared to be sought after by the eldest daughter of the household. She put on some Mexican rock-and-roll and tried to get Mike to dance with her. Unfortunately, Mike's jeans had long ago worn out the seat and he was a smiling wall flower. Bill mentioned to our hosts that his birthday was the next day. That remark brought out the house special, Oaxaca mezcal (Gusano Rojo), or the magical cactus juice with the red worm. We four Americanos were already almost beered to the limit with Tres Equis (Triple XXX beers). It is a shame this beer has been discontinued. It gave a tremendous buzz. Yet, it held no competition for the mezcal.

The difference between tequila and mezcal isn't what we'd initially thought. I'd been under the suspicion that mezcal was made with mescaline cactus that are supposed to induce great visuals. Mexican liquor laws require tequila to be made only in five states and only with blue agave cactus. Mezcal can be made in nine states with a variety of cactuses. Present-day Wikipedia states that Oaxaca cactus juice is made with agave espadin, the most common agave found in Oaxaca.

The petrol taste of the mezcal vanished around the third, gagging shot. Once down the throat, it continued to burn and burn. It was kind of neat that each bottle (we finished two among the group) had a small, red bag tied around the neck that contained pepper and salt. We didn't need the salt, just the encouragement from our new friends of San Juan de Los Cues. We got to know this liquor as 'The Devil's Brew'.

It doesn't matter if we now know the booze isn't made from the hallucinogenic cactus: back then the high was ridiculous, and totally different from tequila or whiskey. I couldn't see anything except blurs. Our eyes might have been high, but my brain didn't seem to be impaired. But isn't that how a drinker rationalizes when they're over their limit?

Birthday Bill was the new love of the village. A small, smiley, spaced-tooth Mexican lifted Bill from the floor and kissed him on both cheeks. When I returned from getting my camera, Birthday Bill was twisting the

ROAD TRIP: HUAUTLA

night away with the farmer's daughter. Mike had wandered off and wasn't available for the photo.

San Juan Los Cues - Bill's Birthday Party

The party could have gone one of two ways, pass out or retch. Sally and I decided we'd had enough liquor and staggered through the town, finally wandering into the cemetery. The plot of tombstones deposited us into a scene from 'Easy Rider'. I was too plowed to focus, but could I see shadows darting between the vaults. The ghosts of San Juan de Los Cues had found us and we were enthralled. Out of nowhere, we were suddenly encircled by a crowd of boys from the village. I trusted my intuition when it told me these guys were super horny. What do you say to the home-town gang when you're so drunk you have a hard time moving your mouth?

Every boy in Mexico has the right to be horny, since they live with the tradition of the virgin bride. Mexican women don't seem to be too hung up on sex. They dress attractively, but not seductively, like in the States. Never do you see a braless woman unless it is chow time for the bebé. Sally wore a solo tank top. There was only one tactic to stagger back to town with the guys pawing at Sal, which wasn't humorless or avoidable. Closer to the village they disappeared, probably afraid of their elders.

It was night, and the game was to find our trashed friends. Mike was rattling on with the host family —even though he knew no Spanish. Bill was in the van, zonked, the first victim of the mescal overdose. Sally eased over to the town store that had everything and was also the local night spot. The party continued there. The older guys impressed upon us how they welcomed our visit. For us, it had been a welcome break in the hot, dusty-road monotony. They were honored when we gave them our addresses and shook hands. One drunk, a definitely important white shirt, scribbled on a piece of paper. He gave it to Sally and said that if we ever needed any help, no matter where, we could just call that number. San Juan de Los Cues would send help. We were now children of the village. We were sincerely honored.

We became members of San Juan de Los Cues.

Stopping for a cold beer had held us up for six hours. This had been the first town where no one tried to hustle us. We felt like out-of-state relatives returning home, instead of tourists expected to drop some money. The moon was one day shy of full and illuminated the plain as the VW roared along the widened horse path to Huautla.

FINALLY HUAUTLA

Sleeping was an adventure in discovery when drunk on mezcal. The van parked in the open desert to avoid smashing an unlucky burro or greeting a cactus head first. Our alarm was a group of farmers riding to their crops just after sun up. They found us hilarious for sleeping in the van. Throughout Mexico, it was our duty to provide morning humor for the locals. They'd watch us and giggle. In the mornings, they would sneak up and peek into the steamy windows. They must have never enjoyed privacy or slept late.

Once our alarm group helped us greet another scorcher, we pieced together the previous evening. We were lucky to find everything we'd believed lost, even Bill's spectacles, and then enjoyed breakfast. We were on the hongos trail. Next stop Teotitlán.

The next town didn't have the character of San Juan. It was a dusty hole directly beneath the mountains. If it hadn't been for the bus stop it probably would not have existed. Teotitlán was a state border town and held the last cluster of people until we reached Huautla. According to the map, this was the most direct road to the mushroom cult. You must travel to Teotitlán in Oaxaca, then pass into Pueblo State and then back into Oaxaca. Government probably set it up that way so they could legally have plenty of guards on the road to the hongos.

It was still early when we reached the special casa fiscal (the border tax station). The guards were a comedy team, one Pancho Villa swing belly and his only soldier. It was their duty to checkout who was traveling to Huautla and get in on the action.

Why were we going to Huautla? For photographs, of course. No hongos, eh? No, what's hongos? It was then the captain hit us up for a joint. We didn't have one, but some hard tack candy was a decent substitute. A local teenage boy was waiting for the bus, so we offered to trade a ride for a guide. We were getting closer. It's pretty hard to get lost when there is only one road.

Outstanding is one word to describe the climb from Teotitlán into the mountains. The road was actually two lanes wide in some places. The big holes and long ruts compared to our route through the desert. Mile for mile, they were equal in terms of rocks and holes. Of course, this mountain highway had the danger of a precipitous drop if you eased too close to the nonexistent berm. Even though it was slow going, there were smiles all around. It may sound corny, but the border station now seemed like the gate to the Garden of Eden. After about an hour, deep green became the color of the day. The road from Teotitlán began construction in 1922 and didn't finish until 1959. Considering that it must have been a horse and pack-animal trail before, our trip was luxurious.

Mexico Mountain View

Deserts are okay if you are a masochist or looking for great answers, but I'd rather enjoy the cool, fresh mountain air. There were no houses close to the road. Some spots were cut into the hillside and had a traditional thatched house. It was great not to see cactus or sugar cane. Every mile it seemed another mountain stream arched onto the road. After two days in the dry heat, a cold shower was a treat.

Our passenger/guide couldn't understand our journey. He'd lived in the village all of his sixteen years and didn't find it very fascinating. He definitely didn't feel it was worth answering a barrage of questions. He said we'd be there in six hours. Again our map varied from reality. This was another unimproved road.

The road seemed it would continue forever; the guide rattled on about the village, psychedelic mushrooms, and strong weed. As most young dudes did, he knew where to get it all—kilos, too. We reinforced our Yankee imperialism with peanut butter and jelly sandwiches and cigarettes.

By noon, we reached the midpoint, two houses on a ridge. It was just a road house and seemed like a good place to soak in some local atmosphere. The food was down-home, all the way. The woman offered two choices, chicken with chili peppers or beans and chili peppers. It was fantastic, hot and spicy. Her kitchen was modern yet primitive, a concrete platform built on to a dirt floor. One side had a fire pit with a grate over it and the other held greasy water. The condition of Mexican restaurants didn't bum us out, because they were right in your face with it. You saw the food being cooked. Of course, sometimes you were going to get 'the revenge,' and you might as well be prepared. Luckily, the van was stocked with anti-diarrhea Lomotil. This chow hit all the spots and burnt on and on.

Mexican Roadhouse Kitchen

Past the road house, the road seemed to level off as the vegetation changed. Before lunch, it had been a steep climb with big trees. Now, it was lush vegetation. It seemed we were in the clouds. I checked later and the elevation was about six thousand feet. Our guide kept saying onward. More of the valleys were cultivated with maize, but seldom did we see any farmers. It was a beautiful day, but there was no one around. Finally, our guide got it across to us that it was a religious festival, The Feast of Assumption.

We were trucking into Huautla, the Mecca of drugdom, on a holy day, under a full moon, and on Bill's birthday. The cosmic entities were combined. Finally, Huautla appeared off to the right in the distance. It didn't seem two hours distant, but it was. Our surroundings were prime, high mountain jungle. It was very different from what we'd experienced in the Yucatán. After all, we'd been on a pilgrimage. I'd wondered where the Indians were going to find banana leaves to wrap the hongos. But there they were: bananas, bamboo, and bright flowers everywhere, along with clean air, and fresh water. Yep, it was a real Shangri-La. If I had known it was going to be that beautiful, I'd gladly have driven another week to get there.

Huautla de Jimenez – Latitude 18.13083 North   Longitude 96.84367 West

Rambling down the road were some Indians returning from tending their crops. Their clothing included unique, embroidered shirts in the brightest colors. Seems these should have been faded, but they had an intense purity of hues.

Huautla rests on the proverbial mountain side and then the road seemed to go no further. With a couple thousand residents (we discovered it had 30,000), it is big for being six hours into the mountains. Our guide departed at a bridge before we reached the city limits. He'd agreed to put together some stuff for us when we came back the following day. We couldn't count on it, though, as everyone loves to brag about what they can put their hands on. The only thing we had going for us were the almighty green backs.

The road wound upward from a raging stream. Huautla was exactly what we hadn't expected: it was commercial. The very special fiscal station was unmanned, but the sign told the story. If we wanted to stay, it would cost us fifty pesos for three days. It was agreed, that we'd be in and out real quick. We did not want to pay for the right just to be there. Monique had advised: score, gobble, split.

Mexico Special Road Tax Station

According to Wikipedia, the Mexican Army positioned a checkpoint on the Huautla Road from 1969 until 1976 to either curtail or profit from foreigners traveling to experience the magic mushrooms. I do not believe the special Casa Fiscal we skirted on our entry was a military station. The ruling Huautla Mexican regime was getting their cut off the top from all the 'shroom touristas. The Special Road Tax sign explained it was for road conservation.

We parked out in the open, near a school, so we wouldn't get our stuff ripped off. The village was laid out with doglegs running east and west. The houses were on the hillside between the roads. At the elbows were the central plaza, the market, the school, and the church. Huautla expanded in the early 1960s with the construction of a municipal building and a high school. Were they financed by hongos? The place looked like a typical Mexican village—peeling paint on cracked plaster—but it was concrete block with plaster and not adobe. Children surrounded the van with the same old, "give me, give me, give me."

Bill and Mike took off. We hadn't paid the entrance fee at the Casa Fiscal and believed the Federales were looking for us. It was a matter of scoring before they found us. Being inconspicuous in a remote Indian village was impossible.

It was then that Huautla became unusual. One of the most vocal little boys wearing a bright orange shirt, pointed to his chest pocket. There was a picture of mushrooms, hand drawn in black ink, and labeled 'hongos'. We nodded and the boy introduced himself with one word, 'Mario'. His mother sold the shirts. We followed him to a fairly new wood house with the word 'sastreria' (seamstress) stenciled next to the door.

Mario slipped into a pathway between the house and a blackened, greasy, cooking shack. Never doubting that a 'high' was available, we followed—almost expecting to be shanghaied at any moment by the ghost of Montezuma. The room had a dirt floor, but compared to what we had seen, in the daylight it was almost cheerful. Where Mario led us housed a ripe, pregnant woman zonked out on the corner bed. One niche had a pile of embroidery remnants. On the table were a few of the infamous Mexican wrestling comics. The walls were bare except for a Banco de Mexico calendar.

Mario's mother entered and began pitching her embroidered shirts. The blouse was stitched with little mushrooms, but we wanted the real thing. After a minute of translation, mamá produced a handful of dried fungi wrapped in newspaper. She pointed an explanation that the woman on the bed was tripping on hongos. Perfectly timed, the pregnant woman twisted on the bed, grinned and whispered "hongos", then slid back into never-never land. I've always wondered how that child turned out.

Arriving at the dollar figure, the woman sighed, explaining that she only had one live hongos, but the dried would keep for smoking or eating. She pressed the fresh one into my mouth as she grabbed the money.

Later, we discovered we had been duped by an elaborate mother-son con game. Mario would attract the tourists and before anyone could state what a reasonable price was, caught in a rush, mamá would bilk them for quadruple the regular going rate. She also sold a local moonshine terror called 'Conan'. We'd been had, but didn't realize it. Sally and I returned to the VW and deposited our treasure. With no sign of our friends, we headed into town central to try and get a better assessment of Huautla.

Mike and Billy also had quite a time scoring 'shrooms. They encountered an old Indian who led them to the most elevated streets of the village. Huautla is very deceptive. Due to the steepness of the hill, seldom can you view more than fifty feet ahead. The streets zigzag and weave, crisscrossing into a tricky maze. The more Sally and I walked, Huautla expanded. My expectations had been one main rutty street with two rows of about twenty huts. Huautla looked more like a small city. The majority of the buildings were concrete. Almost every necessity could be purchased in the one block, termed the 'market district'—shoes, food, farm tools, liquor…anything.

Disappointed that no one else had tried to sell anymore 'get-high,' we strolled to the edge of town. We were just about to turn back when a young Indian man emerged from a house. Si, we wanted hongos.

Hernando was maybe eighteen and seemed to be smart. Most rural Indians felt education didn't put food into their mouths. He'd had some schooling and was aware of what was happening in the world. He actually mentioned Watergate. His adobe hut was fairly together. The cracks between wall boards that didn't quite fit flush were stuffed with newspapers. Just off the doorway was a typical Indian bed, wood frame covered with a woven straw mat. Along a wall was a row of adobe bricks waiting for someone to get the energy to mortar them into place. The far wall had only a plain table and chair. No kitchen was visible. The last wall had the Virgin of Guadalupe beaming out salvation.

He happened to know where to cop some smoke and 'shrooms. Triple A couldn't have directed us better. From his brother's casa next door, Hernando delivered a cup of coffee to Sally and some weed. We were never introduced to the brother, as Hernando would speak into the doorway and they would hand him things. The weed looked and smelled like dynamite, but again, we were stuck without papers. Due to government paranoia, Mexicans don't market any cigarette papers. Bring your own as they roll with brown paper from shopping bags or dump the tobacco from cigarettes. Either causes a sore throat. It didn't matter because there wasn't going to be anything smoked in that house. The Federales would shake down the locals if they caught them dealing, and really sock it to gringos.

When Sally finished the coffee, Hernando led us to another house for mushrooms. The second hongos outlet must have done better business since the house was in much better condition. The small yard was scattered with kids and chickens. Our guide whispered to one child who entered the house. A stately middle-aged Indian woman appeared. Most Indian women are short, at about five feet, but this lady was our height, at five-eight. Her dress was spotless and when she smiled, we got a glimpse of her gold incisor. Gold teeth define the wealthy Indians. Bad teeth describe the poor. All Mexicans seem to have dental plates and it appears to be a prerequisite of marriage that every girl gets her teeth straightened.

This hongos lady would only converse with Hernando in the local Mazatec Indian dialect. It is difficult to say if it was the rule or the custom of Mazatec women never to speak Spanish. The language sounded oriental, but then I'm a big fan of Charlie Chan movies. After a short discussion, the woman reappeared with the infamous banana leaf.

ROAD TRIP: HUAUTLA

These hongos were unlike any mushrooms I'd ever seen. Each resembled a circumcised penis with a white shaft and a black crown. On the leaf, were maybe twenty 'shrooms—probably enough for the two of us to catch a buzz. All for only two dollars. It was then that I felt the first woman's hand in my pocket.

Huautla Mushrooms

After the exchange was completed to everyone's satisfaction, the woman displayed some of her embroidered shirts. They had an amazing blend of colors with an intricately stitched pattern. There was little doubt where she had seen the visuals. We bid her adios and told Hernando we'd check him later. It was 3:00 pm and time to leave Huautla.

On our return, we found Mike stumbling around the church yard wearing one of the embroidered shirts. He'd obviously discovered the local hongos. Bill was OD'd, flat on his back in the van, staring out the open sunroof. Realizing their condition, it seemed to be a wise plan to exit town before we gobbled our 'shrooms, to avoid any delicate situations. There was no recommended dosage, and no idea about duration. An isolated launch pad was a good choice.

BILL AND MIKE'S TRIP

The two gringos sauntered along an empty main street, attracted to an unusual building. Every other structure of the village was a basic raw wood hut or a concrete, two-story painted solid red, blue, or green. Yet, the plastered wall they had been drawn to was a pale yellow stucco wearing a complex dark blue Indian painting. An unfiltered black light was aimed directly at it.

Huautla Store

Leaning against the building's corner was a sleazy Indian wearing a crumpled straw hat. As Bill shuffled by, the man whispered 'Hongos'. Easily seduced, the duo followed him for three blocks until the road merged with a dirt path. The man stopped and requested ten pesos to proceed. To advertise the quality of his hongos, he removed his hat to display a long, unwashed, recent gash in his forehead that had been bleeding. Their guide just pointed to it and again uttered 'Hongos'.

'It will cost you ten pesos,' he said again.

Problem was, Mike only had a fifty and Bill a twenty. Satisfied with the gash, Bill forked over his money and received no change. The Indian led on through a cornfield, passing family homes. In about fifteen huts from the nearest major path, the two were deposited into a doorway.

Inside the home, a middle-aged couple, their adolescent daughter, pre-school son, and a toddler were enjoying a siesta reclining on the dirt floor. While the guide chatted with the family, the young girl efficiently huckstered one of her embroidered hongos shirts to Mike for forty-five pesos. During the transaction, the mother found two large banana leaf packets. Every house must have had a private mushroom garden.

The father laid the packets on the table and motioned to his visitors to come and eat. The fungi, according to Bill, were ordinary and very bitter. The man offered a jar of honey. With the sweet condiment, the hongos were chomped.

As they finished, a young Mexican with long-hair entered the hut. The grinning boy announced he was also tripping on 'shrooms. Fifteen minutes passed and nothing was kicking in, then the mother brought another package of dried mushrooms. About a half-hour later, the guys began noticing distinct alterations in their perceptions. They left the hut and discovered they'd been led to a terraced hillside. No other huts were visible, except on the same level giving a great open view of the scenic valley.

From a lower terrace, there was some commotion. It had been explained that someone was collecting rents from the tenant farmers. Suddenly, the Mexican realtor appeared between two big Indian bodyguards. The three conversed as if the two American hippies were nothing out of the ordinary.

About two hours after the initial munch, the unusual flooded in. The little boy began running back and forth trying to fly a kite in the yard. It was a basic small kite: paper on two sticks. With no tail, it wouldn't elevate. Mike tried to start the kite and realized his equilibrium was altered, weaving among body rushes. He must have impressed them being the friendly, partying Americano as they offered yet another banana leaf packet. This one contained the black and white little prick-like fungi.

A few minutes after the third dose, another young Indian man approached with a yellow school tablet. Fifty was written on it, meaning for the hongos. At the threshold of their high, Bill and Mike realized the first twenty had gone merely for guidance. They decided to haggle for a lower price and offered fifteen pesos. The father realized the guys were arguing and stayed at his fifty. There would be no negotiating. He knew little more Spanish than 'eat' and 'honey' and rattled on in the Indian dialect.

Neither Mike nor Bill could stand upright. The hongos roared on. Bill wrote 'thirty' and the father countered with forty. Mike paid fifty after all the hassle, completely dumbfounding the family who were decent enough to lead them back to Huautla's central plaza. Bill staggered directly to the VW bus and reclined, content to stare at the sky while tripping his ass off. Wanting to explore more, Mike wobbled off in the direction of the church.

A white-shirted man on a bicycle rapped on one of the van's windows and interrupted Bill's 'shroom meditation. The bicycle man initiated the conversation with, "Fifty pesos to stay in town." Incoherently, Bill replied, "Amigos returning and we are splitting." The bicycle man pulled out a ledger with a telegram notice stating that all Mexicans from outside the district, and all hippies, must pay a fifty-peso permit fee.

"A few moments, please, until my amigos come back," Bill muttered. " I will remain here, no keys, I cannot leave."

Official Hongos Telegram

Bill and the hongos reclined on the van's cushions again. The white shirt didn't leave. Suddenly, the church bells began a discordant clamor. The official knocked again, pointed at Bill, and said, "El Presidente."

The only vision Bill had was of being thrown in the local jail while tripping. He struggled out of the van and followed the man across the school yard. The surrounding buildings started to rhumba, but his feet kept plodding forward through the market, into the municipal building and up three flights of stairs. Bill feared he'd lose his balance and topple backward at any moment.

At an office, the white shirt knocked and entered. A few minutes later, a bewildered Bill was called. He faced a huge crew-cut, gray-haired Mexican: El Presidente. The reality was stark; he was the most exaggerated authoritarian figure of all time. It was as if he'd opened a door and been introduced to Hitler.

All the Spanish lessons he'd never taken flashed through his soggy mind. Surely, El Presidente could discern the effects of the hongos? The mayor displayed the federal telegram and inquired about the number of Americans and how long they'd been in Huautla. Finally, pointing to the three on the bicycle man's watch, Bill completed the quiz. The hallucinations were coming on strong. Every angle of the room was swaying and rippling. The pictures on the walls were bouncing. An unbearable, blue glow shined in the window behind El President. Bill could only mumble about his amigos.

The quiz continued. "Why did you come to Huautla? Do you expect to eat hongos? It is illegal!" The mayor reached across his desk for another brown copy of the official telegram. "Huautla must be protected from tourists. The old religion remains and many come to witness without any regard for our town. We have placed a special tariff on them —which your group has disregarded. Now, do you intend to pay the fifty pesos?"

Bill again stuttered that as soon as his amigos returned, they were leaving.

"Time does not matter now. You are in Huautla and must pay or immediately leave. It is very simple," the mayor ordered.

Emptying his peso-less pockets, Bill explained he was broke. He was alone and stoned to the max on 'shrooms. Bill's spirits ebbed. Paradise had been lost forever for a mere fifty pesos. Walking carefully down the stairs, clutching the hand rail, he imagined not stopping at the van. He could walk into the mountainous jungle, alone without clothes or bedding. Where would the official leave him?

Bill remembered the extra ignition key he'd made at the beginning of the trek. Should he desert and return tomorrow for his amigos?

"You must leave, pronto!" shouted the bicycle man.

What a birthday dilemma! He had two choices. One was to drive off to who knows where, and in his drug-induced condition, could he even drive? The other choice would defy El Presidente —and that could only mean trouble.

He decided to delay the action by searching for the wallet he'd hidden when the bus parked in the school yard. Then, he searched through the compartments until he found the hidden key. Ready to go, he gazed into the mirror contemplating his bloodshot eyes. Still woozy, chills ran the course of his spine adding to the intensity of the moment.

Where had everyone vanished? Staring off, gloomy suspicions blotted out the green hillside. Perhaps the Federales had already grabbed the rest of the group. Maybe they were already imprisoned, after all, everyone had planned to check out the town, score, meet back at the van, and split. They'd imagined the possible dangers of the mushroom town. Poor Mike was definitely high and wandering around aimlessly. The negative possibilities of the situation gained control of Bill's enhanced consciousness and persuaded him to take the van and go, before he was most certainly incarcerated.

He started the engine, adjusted the mirrors, and checked the brakes before releasing the hand brake. He was slowly pulling away from the now-grinning bicycle man.

"Hey, hey! Bill, hold up! Where are you going?" The cry came from the direction of the church. His depression vanished. His three companions had returned. They could take the immense burden of the permit off his spaced-out mind. It took a few minutes to explain the situation. The bicycle official flashed the telegram again and I was pleased to part with fifty pesos.

Bill couldn't hide his renewed confidence. Tears welled up behind his specs. "I am so glad to see you guys. This has been a huge hassle, but I'm too high to tell it now." He found his bed in the back of the van and said

NEW EXPERIENCES

Fifty pesos equaled four dollars and wasn't a tremendous sum of money for a three-day visa to paradise. Back then, it was the same price as the toll on the Pennsylvania Turnpike from Pittsburgh to Philadelphia. We rationalized the temporary visa fee wasn't outrageous —but definitely illegal. The immigration official in Mexico City had told us no other visa was necessary. It was easy for Huautla's civil servants to pull it off. In those days, hippies weren't the most honored visitors anywhere. After enduring the torturous bus ride, the officials wouldn't believe any reason other than the sacred mushrooms. If there were any complaints of unauthorized permits from the pilgrims, they would've fallen upon deaf ears.

For only eighteen, Sally was sophisticated. As I drove farther up the road, Bill and Mike incessantly babbled, nothing much coherent. She proceeded to twist up some of Hernando's gift.

Our initial visit to Huautla had been hectic. All we'd done was rush around —first to not pay the small fee, then to find some dope, and then to pay the small fee. Finally, we could relax, confident the mushroom thing was basically legal and we had little to fear from the authorities.

Huautla

Taking the time to pause and reflect on where we were, Huautla was beautiful. It is at the center of a well-cultivated, coffee-producing area. As we drove, we could see farmers in white cotton shirts and trousers with straw hats chopping the unending weeds. The road worsened with huge mud holes. I'd hoped we didn't get stuck —as my saucer-eyed buddies didn't look as though they were in any condition to push. Luckily, we bounced along through every hole. Research tells that the name 'Huautla' is from the Mazatec word, 'Nde'le – place of' and 'Ha - eagles' place of the eagles. I wonder if the mushrooms helped us to become eagles. According to the Mazateco Newspaper, Huautla began around 700 CE when the site was founded by the twelve tribes. With the influence of Catholicism, the village became 'San Juan Evangelista', after Apostle John. It gained 'de Jiménez' from José Maríano Jiménez, a general and hero in the Mexican war of Independence who was executed by a firing squad in 1811.

Our route took us to the opposing hillside which faced Huautla. Our ride continued until I found enough space to safely pull off the road. We hadn't seen any other cars, but if we were going to dose ourselves, a good parking spot was necessary. A sign written in English warned against littering or bathing in the mountain streams. That certified how many other freaks had proceeded to that point. It marked the boundary of some -'Juans' coffee plantation.

Another stream tumbled off the hillside. The region was a tremendous watershed. With no one living above, the water should have been drinkable. Finally, there was a clearing directly facing Huautla that was big enough for the van. Hernando's weed had been good, but Sally and I definitely weren't anywhere near Bill and Mike's 'back-in-a-minute stares'.

Hongos Close up

Following a decent parking job, and careful not to careen off the precipice into the valley a thousand meters below, I unwrapped the banana bundle. If they really were mushrooms, they surpassed anything we'd encountered in our previous escapades. This should be a once-in-a-lifetime experience. I did a total check out: no smell, smooth texture, and no sounds. I think I expected a hum. After close-up pictures had been taken, it was time to enjoy the fabled hongos. The first bite was almost revolting. It was so acidic, nothing except puckering juice. These mushrooms were completely different from the firm tasteless variety that accompanied US steaks. It is hard to expect something grown in cow shit to be palatable.

Sally wasn't eager to enjoy the total abandonment of the boys in the back. After a bit of prodding she chewed a few. I consumed only half of the banana packet and she a quarter. The California freak brothers advised two packets, but from the appearance of Bill and Mike, this was some powerful stuff.

Mike was getting it together and decided to explore our parking spot. Bill was still flat on his back. Neither could estimate how long it took them to get off. We had no idea what to expect. We again put the weed to good use, ignoring the warning from Los Cues.

A nearby stream offered me a chance to bathe and get off by myself. I decided to climb up the hillside a bit so that when I felt 'shrooms coming on, a return to the van wouldn't be difficult. The hill was steep. A bit of isolation felt good. About 200 feet above the van, I stopped to just enjoy the scene of evening approaching the valley. The clouds swept across Huautla. I was in the middle of about a two-acre cornfield split by a stream that produced the only sounds. Every ten minutes, a farmer would pass nodding his head. They were returning home before sunset and I was on a main artery path between the terraces.

I measured my psychic climb by how I conducted these meetings with the farmers. Between the third and fourth "Como esta," the corn stalks began to vibrate —shaking their tassels in choreographed unison. I noticed the rhythm of the breeze, the waterfalls, and the crickets as they welcomed the night. The clouds were no longer pure white, but had prismatic rainbows highlighting their fluffs. I was feeling fine. The silence had tuned me into nature.

The 'shrooms started to roar through my nervous system. Euphoria took over. The lush green corn began to radiate and wave while the puffy clouds seemed to pulse as though they were pumping to the rhythm of the waterfalls. One farmer in a dusty white shirt and trousers, riding on a burro, stopped to converse with a loco Americano. He had bare, calloused feet and brown, knuckled hands. My gaze elevated until I stared into his deep brown face, trimmed with gray hair and stubble. He lit a cigarette and puffed out the smoke. I looked into his eyes. My energy was vibrating. All I could mumble was, "Huautla is very beautiful."

The farmer agreed and explained that the coffee trees, sky, water, and people were gifts of God. Those elements were all that was necessary to experience happiness. Smiling into his deep brown eyes I exclaimed that I was happy. The farmer bent and picked up a handful of the dark soil. He proclaimed it the best in all the country.

By now, my nervous system was experiencing a brown-out. As we parted, I gazed at the van below and noticed Sally was having an animated conversation with young Mexican, maybe twenty-years old, while Mike and two farmers looked on. One of the farmers had been watching from below the road as we parked. Even in my present condition, I could sense some bad vibes in the evening air. My high slightly waned as I noticed tenseness in Sally's expression.

Huautla Farmer

Getting back to the van was a stumble-bum mission, but I returned in time to catch the punch line that we owed him for permission to park. With a gluttonous stomach stretching a black jersey, he looked the part of the bad guy in this scene. The three locals swayed as if they'd had their fill of liquor. Their goal was to frighten and extort some cash from tourists. He was definitely the cool one, in pointy shoes and shiny black pants, laughing and snickering with his buddies.

Sally explained that we'd just married and the van was our only home. We didn't have much extra cash. The punk didn't buy it and insisted we pay four dollars. These small-time extortionists were sure we were touristas — loaded with dinero and quick to pay to avoid hassles. It was insulting to be plagued by these slimy characters.

Some very strange feelings were rolling up my spine when he asked if we'd eaten hongos yet. We were stoned and it should have been obvious. The sleazy dude stared into my eyes, which probably defined the moment. Mushrooms have their telltale signs, like profuse sweating and a non-focused stare. With all other drugs I'd experienced, my pupils oscillated —yet, with this natural psilocybin, they were dilating to a pin-point. It was as if this negative force had triggered a system failure and I crumpled to my knees.

In the split second before I passed out, I remember looking at the hillside above the three Mexicans. Suddenly, a bright orange rectangle materialized in the center of my view, totally blocking out the tree tops. It became a photograph with an orange doorway super imposed. As I sank to me knees I truly felt my spirit drawn through the orange opening. My mind detached from my collapsing body, which was now useless. My thought process seemed unaffected. I could still think logically. My system suffered no fright from the magnetism of the orange rectangle. Looking back, just before my being passed into the rectangle, I witnessed my own crumpled body, supported by Sally and Mike. Suddenly, that consciousness was gone.

I transcended beyond the orange warp. It was a world, dimension, or something-someplace that consisted of very bright hues and geometric configurations. My self no longer had any form —only a glow I could identify as 'me'. I don't even know how I would have seen it, except I knew I was a glow. Wherever I was, it was absent of depth and there was no bottom or gravity, only color intersected by lines. A description is difficult; words haven't yet been created to present a distinct picture. The best available analogy would be looking through a lighted prism. Each angle of the crystal refracted a range of color and light. I sort of floated among the angles of bright light.

Time had been abandoned on the other side of the orange door. Time was present though, because things happened; they started, ended, and something else started again. I could comprehend intervals. Time being equated to different shades of the day had vanished. I lost the conception as I tried to explore the extra consciousness. I was enjoying the view and I fully understood I was 'someplace else'. Another glow joined me and welcomed me in a language I understood. There was no accent.

"We are glad you entered." I'll never forget those words. "All questioners come here. You asked this question. We and this are the answer."

The glow guided me to other glows —who in-turn, issued a welcome. It seemed they recognized me, or something familiar in my glow. Like I said, I couldn't see my own glow. Maybe I had a license plate or something. A glow approached (floated or something propelled it) from a brilliant green area and saluted me with something that sounded like at least a five syllable name. It sounded something eastern or Indian. I didn't retain it. That glow conversed with me for a longer period. I can't remember the exact words except that the inference was more than familiarity. It was as if we knew or had known each other. More glows surrounded me and I became the center of a conversation that dealt with life. Not so much living, in my world or theirs, but instead about the life force. They repeated that they'd known I would be with them again.

Suddenly I heard another voice calling me. The intensely hued world disappeared as I awoke in Sally's arms.

'Are you all right?' Focusing was difficult, but once I returned to this world, Sally and Mike looked as if I'd been resurrected. 'Wow, we were worried about you'.

The Mexicans had been so shocked by my faint, they'd moved down the road about a hundred meters. There were zero reasons for them to not take advantage of such a pushover situation. The three land sharks sat and watched us as if waiting until nightfall for a greater chance. I passed out again and guessed that my amigos had left me to lie on the ground.

Regaining a spec of consciousness, I realized I was soaking wet —as if my body had undergone some strange convulsion. My wobbly knees would barely support me. All I could do was lean against the van. I checked my appearance in the side mirror. My lips were bloodless and my face frighteningly pale. A chill ran through me and I removed my jersey, actually wringing water from it.

No one else was around the van. With effort, I retrieved a blanket from the van and lay down on the front of the overlook. I succumbed again, maybe for another fifteen minutes—a very memorable fifteen. It wasn't unconsciousness, but something else. Something very nice and fun.

This time, there was no orange doorway. I was surrounded by glows as I'd been before I'd been awakened. This instance, the conversation, which before had been muddled, I could hear clearly. It was a uniform chant: "We are within; you are within."

"Within what?" came from my glow.

"Your being," was the chorus' answer.

The glows receded—except one who became my guide or mentor. "The route you've taken has been designed to reveal this world. Forever, they have worshipped this pathway to the spirit. What happens is the traveler sees within himself, deep within. This world belongs to everyone. You've barely seen a grain of all that exists here —since everything exists here." The guide relented. We were moving, drifting over geometric shapes.

"What do you want to know?"

I wondered what questions to ask. My mind wouldn't respond. Everything that had perturbed me was gone; things were so simple. How to be happy? I was. Money? No need. Religion? I was experiencing. Politics? Who cared?

"This world truly exists as much as the one you exited. Contact by the means you've used leaves the being short. You were destined to come full across, witness, and taste."

My senses felt I didn't know as much as this glow expected. I was missing the point. It was above my pay scale. I thought about whether the glow had a sex. As the question formed, the glow answered.

"I am neither and both. Several times I've spanned the dimensions, sometimes as an object. It is not of my choosing. Each experience offers education. Ours, as yours, must be to learn; otherwise, you would not be here. My being is dedicated to perception. My spirit would not be here if less inquisitive. Without asking, I have learned from your life's book and gained from it. We enjoy constant information from visitors of other states of being. We never anticipate transcendence to another level."

Understanding, only in a fleeting instant, what was happening, I felt isolated. I wasn't into the head as much as I thought I could have been — and that was holding me back. Maybe I could have been more active in the role, and exploring it, but it was easier to permit the other being to lead. Maybe I was permitting? It was a great answer, and I didn't even have to ask a question; something flittered into my thoughts and the other glow responded. It occurred that I could concentrate and unleash something inside. I could become one with the ancients. Where did that come from? Was I an ancient? But I couldn't completely let go. There was the fear of losing control. My surroundings began to accelerate, but my thoughts were not coming quicker. The background shapes glided by. What could I do? Hesitation became the break. I could sense descent and weight. I awoke on the ground, wrapped in a blanket. My clothes were damp with perspiration. I was alone.

It still wasn't quite night. A few thin yellow lines separated the clouds and the horizon. The three Mexicans rip-offs were keeping a vigil. I called out and no one answered. With night descending and two friends high, I expected the worst. I beeped the horn until I realized I was a victim of paranoia.

Again on the ground, I began to let the vivid hallucinations amuse me, the spectator of illusions. Vivid patterns grew from a single star. The clouds became paisleys and more paisleys, until I was sitting inside a complete paisley atmosphere. The chill revived my mind again. The stars were in full bloom. The moon wasn't very far up yet and the amazing celestial patterns radiated.

"You have your act together yet?" My two wandering gringo amigos had returned, signaling the beginning of this play's second act. "We heard the horn and finally realized somebody might be worried."

"How are the hongos treating you, Sally? Did you get off yet? Maybe you should take the rest?" I offered.

"Can't say that I'm not off; but nowhere near the effect the mushrooms had on you. Wow, the way you looked when you went under. Your eyes rolled like you were gonna pass."

"We should have better prepared for the night and had some firewood. This area has been picked clean. Damn, it is getting cold." Mike chimed.

Huautla is several hundred miles south of the Tropic of Cancer, but it rested at quite an altitude, almost a mile high. Huautla is on the eastern edge of the Sierra Madres. The winds coming from both the western Pacific and the eastern Gulf of Mexico poured their contents into these mountains. Every air current is loaded with moisture. The latitude kept the mountains warm in the daylight. The friction between the air currents and variable temperatures produced static electricity.

That night, we were entertained by blasts of lightning along the mountain tops. It wasn't streaks of lightening, and never any thunder, but bursts of light—like a cloud of electricity had blown up. That energy, coupled with the hongos, must have produced a meaningful religious aurora for this region. The members of the Mazatec tribe were the earthly caretakers for all who resided in the valley, both living and spiritual.

Perhaps in another location, the young rip-off guy could have turned that evening into a bummer. The hongos' effect was very strong. Anywhere else and it might have been an overdose. But Huautla is a paradise and I experienced no inner terror. There were no intrusions by passing airplanes or automobiles. No extra people wandering by, except for my companions. The futility of our modern society was what usually crept into my stateside highs. Huautla was basic. This village explained civilized man's society in uncomplicated terms. At one end of this valley, was a town that took every advantage of travelers who had journeyed to experience a remarkable, natural gift. Everyone we'd met had had their special, pocket-filling agenda. They had trained their children to be hustlers, plainly over-charged, short-changed us, and rudely attempted to extort us.

At this opposite end of the valley, the farmers I'd met appreciated nature's own beauty. They were primitive and not distracted by anything except their crops. Money meant survival, not extravagance or luxury. They slept on dirt floors of mud huts, usually with their animals. Their bedding and serapes were woven from their lambs' wool. They ate their crops, bartered, and cooked everything over a fireside in the corner of their hut. They kept their lives tied to the earth. Their problems were few while their joys were simple, but plentiful.

The hongos had plugged me into something. New perspectives came flooding in. Hallucinating wildly, I could make out Bill and Mike wobbling closer. Sally had passed out under the influence, directly onto her face. We decided to let her naturally awaken. The time spent under would be the experience. As we sobered, I babbled about what I'd witnessed.

A half hour passed before Sally stirred with a swollen nose. She and I compared our notes; our voyages were amazingly similar. She had seen the orange rectangle and transcended, remembering identical aspects. Even the patterns our minds had created had startling similarities.

My companions slept inside the van. I chose the ground under the stars. It was a full moon party on Bill's birthday. Quite a party!

MAIN STREET HUAUTLA

The next morning was undeniably refreshing with a cool breeze blowing and a nice temperature in the high seventies. All four of our spirits were totally rested. The local Indians hadn't awakened us. They got to their fields at sunrise and cranked until noon, probably coffeed up. About 1pm, it seemed, they started drinking alcohol.

We decided to make use of our visitors' permit. Main Street became the backdrop, with Garcia the merchant and Lopez the shoemaker. Garcia, by coincidence, was the brother of El Presidente. One store, with a vivid pink front, caught our attention. It was a bit unusual, selling mushroom paraphernalia. The inside walls were racked with black-and-white photos of Indian women and the famous psychedelic mushrooms. Shirts and 'shroom books were on shelves. This family catered to the tremendous tourist trade, (Huautla was the place to go that very few knew about), a tourist trade that we hadn't realized. This store, and a greasy taco parlor down the street, were evidence that a few capitalists understood what was happening. On one wall, hung the pattern that I was certain I'd viewed under the influence. A black triangle, with a big, black foot inside, was very distinguishable. Sally agreed. No one could explain why it was on the wall or what it was supposed to mean.

There were photos and postcards of sacred women in their traditional dresses. That was probably all they ever wore. In the pictures, all were carrying large amounts of mushrooms. Some of the wall photos showed the young women who were initiated into being a sacerdosita. There was also a photo of an older man who was a leader in the mushroom cult, but there was no name on it. As time went on, it was very odd no one mentioned a word about him. We sensed he was very powerful and even more isolated.

Note: Sacerdosita was the reverent term used among the villagers to describe a woman who had been initiated as a priestess in the mushroom cult. Recent research identifies sacred women now as 'curandera', which translates as 'healer'.

It was decided we should meet Huautla's El Presidente totally sober. He was the man at the top of Shangri-La's politics. With Sally as a translator, there was none of the usual turmoil that comes with meeting such a VIP. We knocked, entered, and then met perhaps the biggest dog in the Huautla yard. He looked as though he could play American pro-football. Introducing himself, Mayor Garcia was pleased Americans came to meet him without any problems.

ROAD TRIP: HUAUTLA

El Presidente - Mayor Garcia

El Presidente quizzed us as to the reason for our visit. He hoped it wasn't to experiment with the hongos. He showed us the infamous telegram he had under the glass on his desk. On the wall, was a plaque that permitted the Mazatec Indians to legally continue their mushroom religion. The mayor was so honored we wanted his photo; he brought out the city register for us to sign. The book was specifically for foreign visitors and was filled with signatures from Japan, Canada, France, and England. Huautla's unique notoriety was worldwide.

Mexican Government Permit for Mazatecs to Use Hongos.

Interlocked with the Mazatec indigenous mushroom religion is Christianity. Huautla's central cathedral was first constructed in 1766 of traditional mud, stone, and grass. It got its bells a hundred years later. Hermenegildo Ramírez Sánchez was appointed bishop of Huautla by Pope Paul VI on January 4, 1975.

We hit the local market with some money. This was exhilarating as the vendors sold goods with a smile. The selection wasn't large, but some stuff was unique. The entire market was open, with no stalls or booths, only Indian women squatting on their blankets with goods sprawled out around them. Piles of embroidery, sugar cane candy, dried chili peppers, fruit, and beads were offered. One woman hand-squeezed a large soda glass with orange juice for a peso. Oranges were twenty for a peso. Men sold live, bound chickens, pigs, and turkeys. The larger items that needed transport weren't moving very quickly.

In the back row of this market were fly-ridden tables of fresh meat. A freshly butchered bloody cow's ass hung from a sturdy clothes line with rows of sausages. These tables were luckily in the shade. The market lay directly behind the school, municipal building, and the church. Every day, there were vendors, but Thursday was the day when everyone hawked their goods. No matter how far, they carried their goods to and from the market on their backs, lashed with a strap that wrapped around their foreheads. I saw women walking, tilted forward, carrying a bushel of hot peppers, or a load of firewood, with that forehead contraption.

Huautla Market Day

Women wore long, white cotton dresses, intricately embroidered and sewn with lace and purple satin cloth. Exotic doesn't quite describe their attire, maybe extra-terrestrial. Alternating bands of blue and red satin encircled their necks. The bodices had figures, embroidered cross-stitching on every arm and around the midriff. The Mazatec style is sewing designs of stick people. Below the midriff, is another ten inches of red and blue satin. The sack dress continues to the ankles with another band of figures. In my research, I've found these dresses, demonstrative of the specific Mazatec stitch, are called 'huipil'.

(It was also this huipil of the traditional, Huautla style that gave María Sabina her fame. Gordon Wasson promised not to mention María's name in his Life Magazine article and instead credited 'Eva Mendez' as the sacred woman who performed the mushroom ritual in a village he named ' Sierra Mixteca'. A photographer from San Francisco read the Life article and headed to Oaxaca in 1957. The photographer supposedly saw a photo of María Sabina in Oaxaca City and recognized the clothing as that of the sacred woman who tripped Wasson. It was that photographer who unleashed the world upon Huautla and María Sabina—not Wasson.)

Huautla Mother and Daughter Showing Traditional Stitch

We left the market for the sastreria where Sally and I had originally met the hongos woman who still owed me five pesos. I now realized how badly she'd rudely taken advantage of us. She had lovely dresses, but as we bartered, the price shot higher. That woman really knew the value of a dollar.

As we departed her shop, we got the old 'Pssst!' and were hailed across the street. An elderly Indian woman had her great grandson hawking her shirts. This woman's goods were easily recognizable as distinctive originals. She had reproduced the patterns one only saw when flying on hongos. We purchased all of her shirts and then she displayed the dress she was sewing.

Each woman makes a new dress when the present one is wearing thin. We had already given her a small fortune (thirty-two dollars) by Mazatec standards for her work. Her art had made us happy and we rewarded it the good-old American way, with money. Sally smooched the woman's cheek and the seamstress became as giddy as a teenager.

The old woman whispered 'Hongos' and pointed to another woman standing in a darkened doorway. I seemed to still be 'switched on' from the previous evening. The others were showing no signs of wear or tear.

Bill and Mike had become bored with the seamstresses and returned to the market. It was almost noon, closing time for the vendors. The afternoon sun was wilting their vegetables and their fields required some tending before the mandatory siesta. We walked on down the entrance road until we came to the cemetery. Earlier, driving to and from town, we had seen a few grave-diggers' at work. This was identical to the one at San Juan de Los Cues with raised crypts except our vision had been blurred by mezcal. Many of these tombs were painted blue.

Inside the sooty niches on the headstones, where candles burned, were photos of the resident. The stones were tightly packed with no walkways. There was an abundance of smaller tombs of children. The cemetery stretched down a steep slant into the valley. The local population explosion and demise were evident.

We watched as one family laid a member to rest. As the pallbearers and family departed no one seemed depressed at losing a loved one. They seemed more happy than sad. The grave diggers staggered to keep up with the family. Not wanting to be disrespectful, Sally and I watched the funeral from outside the cemetery. No one cried, and no one else seemed touched with remorse. As they left the grave, everyone was back on track with a smile.

Huautla Cemetery

The Mazatecs are called 'The People of the Deer'. As the farmer had described the night before, everyone tried to live in harmony with nature. As with most small, isolated villages, everyone was somehow related. The groups of houses on the terraces held extended families. The women cared for the home and children while the men gardened. Everyone knew everyone who belonged—and it was obvious who didn't. In 1973, no one in Huautla recognized it was on the cusp of change. The village had been remote and almost timeless, but the modern world approached with visitors like us. Eventually, they would want what they didn't need.

Death was only a transition for the Mazatecs. The religion of the hongos provided more than a glimpse of the world beyond. People weren't sad to have a family member die because they knew they would all meet again…on the other side.

We were hours late for our rendezvous with Hernando. Just as we'd made our last turn, we were propositioned by an old Indian man who had definitely seen better days. His white cotton clothes were grimy, straw hat crumpled, and his bare feet were hideously calloused and gnarled. He happened to know where there was a kilo of dried mushrooms. It would be available the next day if we gave him half the money. He quieted as the Hernando's hongos woman walked by. She looked at the old man and shook her head as she strolled with her daughter.

Even though we were four hours late, Hernando hadn't organized anything of an illegal nature. No quantities were available. It seemed everything was either already promised or they didn't have large amounts—only enough to smoke. It was too dangerous with the Mexicans policing the Indian village.

A bit of yesterday's get-high still remained. Instead of a baggie it had been wrapped in a piece of the Oaxaca Times. Rolling the last joint, I realized it was completely manicured—no seeds or even tiny stems. It was dark brown and even blackish, reminiscent of some African hooch I'd encountered. It was definitely some well-cured bush, light and fluffy. Thinking we were leaving Huautla that evening, Sally and I shared our last smoke with Hernando. Wow! This weed was some of the best ever. The hongos the previous evening had been so good, we hadn't noticed the weed.

Good smoke brought on the munchies. Huautla's only restaurant called. The restaurant had no sign and no name, except 'Comida'. It was a Mexican version of a mom-and-pop family eatery. The tables had red checkered table cloths. Eating south of the border was always adventurous. Every little place had its own name for items on the chalk board menu. 'Comida corrida' seems to translate to their 'fast food' or 'food run'. We found it to be the day's special. It can be three to six courses. Soup was always a thin, spicy red broth with noodles. The soup never had a particular flavor, except extremely hot. Its base must have been chili peppers.

We'd encountered fantastic entrees such as spicy chicken livers in mole sauce or spicy fish. In this eatery in Huautla, the main course was a spicy stew of stringy beef with rice. Tortillas and frijoles were always on hand. We paid 80¢ and the locals probably considerably less.

Into the frijoles, Mike and Bill arrived. They had the amazing luck of meeting three American girls traveling with a bilingual Mexican, long haired, college student from Colorado, named Victor. The girls had rented a house for $1.75 a day and we were welcome. According to Victor, the scoop on the hongos cult was that only certain women were permitted by law to dispense the mushrooms. The women had to be sacred, trained from their youth. The girls' landlord was the son of the most renowned, sacred, and eldest hongos woman: María Sabina.

We finished dinner and returned to the van. Bill warned that the street to the house was steep and narrow. He recommended a walk. The Volks had traversed the worst of the road and would survive again. We rumbled into the night, cruising through several turns and finally arriving at the last bend with the steep climb. Suddenly, in the middle of the grade, the van lurched to a complete, unintentional halt. I couldn't imagine the cause, but gear oil was leaking from under the rear of the van. It looked like a crack in the transaxle had ended our jaunt.

We grabbed what was necessary for the night and trudged on. Remarkably, in the darkness, we located the girls' one-room, wood house among many one-room wooden houses. The girls and Victor welcomed us. A cooking fire in the far corner was the only illumination. All appeared Mexican with straight, black hair. Victor was at the fireside with Rosa. The two other ladies were warming themselves against the damp night under a blanket. Introductions were exchanged coupled with Kahlúa and limes. The ladies rolled a couple joints and the party was on.

Victor was excited at a chance to talk with more gringos about Huautla. As the bottle of Kahlúa made a third pass, Victor began explaining the religion of the hongos. First, there were sacred women who were the only ones legally permitted to dispense mushrooms, and only at their residences. Supposedly, only seven women and one man were legally permitted to distribute hongos for money. The postcard we'd seen at the souvenir store must have been those women. The sacerdositas were trained from very young to administer the drug as priestesses. The gold-toothed woman Hernando had introduced was one of them.

It is a big question as to how someone was trained to handle a steady diet of mushrooms. Perhaps they'd been trained to enter the other dimension without needing the hongos. That would be the most valuable aspect to learn, if only you could control the changing of dimensions. Maybe the insane asylums are filled with those who can't control the changes. Victor didn't know the principles behind their preparation, but he did know each woman cultivated their own special 'shroom brand in secret caves.

"I've come to Huautla three times. Each visit was more remarkable than the previous," Victor continued, "I was fifteen and working as a waiter in Acapulco. Some of the other waiters were heading to Huautla for a holiday. They spoke of this village with such enthusiasm that I decided to join. We had smoked some primo weed, but I had no idea what to expect with the hongos. It had to be macho. There were three of us and we rode buses, arriving here late in the evening. It was very strange, eerie, walking from the bus stop to the head sacred woman's house."

"I've been back twice learning more of the intricacies of Huautla's hongos." Victor reported. "There are seven sacred women and several initiates. María Sabina is the oldest and wisest. She is eighty or more years old and lives at the crest of the mountains. Her house views hundreds of square kilometers of lush forests, waterfalls, and billowing clouds. She may be what most would consider a bruja, a witch, but María is good—very, very good. She is a true part of nature. María knows all the fine workings of this world and others. Many would fear putting their spirit in such a woman's hands. I do not. When I first met her, she projected kindness and I felt I could trust her like my mother."

"In my first experience with hongos, I did not understand what happened between my mind and body. María gave me a special, very comfortable shirt to wear. She took complete charge of my spirit when it could no longer remain in my heavy, earth-bound body," he confessed. "The sacerdositas aren't witches in any sense, only mental guardians or guides. I don't know, but you must do it."

'Surely you must have heard about Huautla before you met the two sailors in Tampico? Do you not listen to your own music? Who was the group that sang about the rabbit after eating the mushrooms?" he pondered. "The greatest group of all time was here, The Beatles, after their US concerts in San Francisco. I was told by locals it was Bob Dylan who mentioned to them about Huautla. He came here in the early '60s. Others say it was the Grateful Dead who directed the Beatles here."

This info would have floored us if we hadn't already been on the floor.

"Who is the group with the Bear? Ah, Canned Heat. Yes, they wrote a song about Huautla on one of their albums. Santana was here. The album 'Abraxas' is about Huautla. María herself made recordings of her chants," Victor rattled on. We were too stoned and comfortable with the blankets and the fire to interrupt.

"The last time I was at María's, there was a Jim Morrison poster on the wall. María is very famous. Everyone in this village loves and respects her. She attracted the Rolling Stones. The appreciated Huautla so much they played a concert here. Ask anyone." (We did and it seems that there was no verification. Maybe just local hype or it happened and not one of the foreign notables is saying 'boo' about it. This appears to be local folklore. I added this from Victor because it is local Huautla-legend/hype, which almost every foreign visitor learns. Canned Heat does have a Huautla song on 'Hallelujah', their fourth album released in 1969. My research discovered it was Gordon Wasson and the Life magazine article that attracted Timothy Leary and drastically evolved our culture.)

We were all stunned to learn of the supposed influence Huautla had on our music and poetry. I certainly was not going to pass on the chance of a lifetime's experience of getting it on with the Western Hemisphere's 'queen of highs'. Twelve dollars was not very much to pay for the finest guide to the world of the hongos. My present problem with the van had to get sorted out first so I could clear my head.

María could be one of the most knowledgeable unknowns in the world—and undoubtedly, had unlimited power within the various frequencies of her existence. Yet, she was still a barefoot old woman in a very secluded Indian village. María's talents were well known, even worldwide. According to Victor, María did house calls. In the lower valley was a natural airfield. María, with her special hongos preserved in honey, would travel in by private planes to Mexico City provided by those who could afford the luxury. (This is doubtful, but we heard the story a few times.)

I was all for another experience, especially with the world's finest hongos guide.

VAN MECHANICS

Ruefully, the following day began as a dull, gray, ominous morning. I was awakened at first light by a baby crying. An infant in discomfort couldn't be a good sign. Again, laughing Indians pushed their faces to the van's windows, obviously poking fun.

The van rested on a forty-five degree slope in the middle of the dirt path we'd tried to climb. The two homes we'd stopped between provided a constant flow of inquiring children and drunken Indians. Each house must have had ten kids. Men rode burros carrying rusty water barrels, and each rider was a self-professed mechanic.

I had a decent tool box and removed the rear wheel and brakes to discover bad luck. By 10 am, I realized our bouncing through the desert, and over the pot-holed mountain roads, had bent one of the axels. The axel had worn against its housing until the bearing was ruined. The bearing connected to the axel, the axel connected to the transmission, and we were no longer were connected to a ride back to the USA. The van wasn't moving without a new bearing.

What were the chances of finding a mechanic in Huautla? I needed a special gear puller to remove the busted bearing. There was no mechanic, not even a black-smith. We had our permiso and I envisioned paying to use the mayor's telephone to send for parts. I must have been dreaming.

At that time there were only two telephones in Huautla: one for El Presidente Garcia and one public, long-distance transmitter. Everywhere from Huautla was long distance. The public phone was only operational after the municipal office closed so there would be no interference with urgent official business.

The mayor was full of answers to questions from four unwashed and slightly greasy Americanos. No, there were mechanics. No, we couldn't use his telephone. No, he wouldn't call and ask to have the parts sent in my name—even if we paid him.

There was a mechanic in Teotitlán and the mayor thought he should have the necessary parts. We had to hurry because the bus left in twenty minutes. The walk from the van to the municipal building was at least a mile. The mayor's brother, the owner of the general store, would cash traveler's checks for a dollar fee on every ten. The Garcia brothers' hands were folded when they weren't rubbing them together. They weren't often folded.

Bill and Mike had had enough of bus rides and decided to stay with the van, the three girls, and Victor. I was off to Teotitlán in search of parts. Sally agreed to accompany as translator and companion. Her company, steeped in youthful optimism, brightened the journey.

We'd watched several buses bouncing along the mountain road and prayed we'd never have to experience that ride. This was an ordinary school bus scheduled for a twice-daily trip for workers and visitors. This bus was the only regular source of external communication, other than the two previously mentioned phones. All mail came on the bus.

It had never occurred to me that Indians commuted for work. They'd ride for a few miles and get off in the middle of nowhere and someone else would get on. The six-hour ride, with only one scheduled stop, wasn't expensive, and less than a dollar to the next semblance of civilization. Travelers packed their belongings into cardboard boxes and secured them with rope. Once aboard, the boxes were stored on luggage racks above the seats. Every bump we hit (and we hit hundreds) the boxes were hurled at the riders.

We happened to sit among some Mexican freaks that had tripped with a sacred woman the previous night. They were still pretty ripped and bitching about the steep price. Cigarettes were in demand as there was little to do except look at the passing scenery.

The bus halted at the midway road house. It has somehow lost its innate beauty with two dozen unruly passengers squawking, 'Gimme, gimme, gimme'. The beautiful view was obscured by a half-dozen Mexicans pissing at the roadside.

The bus met Teotitlán at five that afternoon. We'd already explored it when we passed through previously and this time, I tried to see even less. The obese mechanic was tinkering with a '59 Ford station wagon, up on blocks. He was so huge that it was possible he didn't need a hydraulic jack to lift the cars. His size was equaled by his grease. No parts. Sure, next town, Tehuacán, would have them.

My God! Tehuacán had a Volkswagen dealership!

An automobile new-car dealer was religiously respected by the locals. The three Mexican mechanics leaned on separate poles of the tripod they were using to pull an engine. Each assured and reassured the Tehuacán dealer had more parts than Germany. Sorry, though, no lend tools.

This trip was rapidly descending into the 'really too bad' stage. We'd left the Oaxaca state capital Tuesday morning and cruised that day in the hot sun. Wednesday, it was the mountains and hongos' craziness. Thursday, it was mechanics. Friday had no aid in sight. Saturday, the car dealership was only open until noon.

We ran to catch the last bus out of Teotitlán at six. Buses only ran either to Tehuacán or to Huautla. This bus was Greyhound style. First class was a dollar and a quarter for an hour of leaving the driving to an aspiring Indy 500 chauffeur. The terrain was flat, the road good, and the driver's foot heavy on the diesel. We roared into the twilight and passed everything: horses, cows, and wagons. It felt reassuring there would be no driving at night. Mexican buses appeared on the horizon and passed minutes later, hauling ass. It was rumored the buses turned off their lights at night to see the road clearer. We'd seen at least one who hadn't properly negotiated a bend and had slid to a stop on its side.

Mexican Bus Overturned

Horse and wagons didn't usually have lights or reflectors, nor did cows or people. Mexicans in rural areas didn't follow any rules of the road for walking. They just didn't seem to care. They never stood in any organized lines for anything. Each waved his pesos, struggled closely as possible to the intended target, and shouted until they were noticed. Riding buses, they'd cramp into small spaces and even hang on to the outside, standing on the rear bumper clinging to the exhaust stack for a free ride. Sally had told us she'd once stood for twelve hours on a bus ride through the Baja. The downside was people carried their produce and livestock to and from the market on these buses. They stood holding the chickens and turkeys by their bound legs.

Times were becoming more difficult. Revenge cramps hit me just as we reached the suburbs of Tehuacán. Diarrhea is no fun, but this was intense pain. Where do you find relief on a bus? It was time to grit teeth! I'd had it before, but the pills were back in the VW.

The runs are inescapable in Mexico. Everyone gets it, some sooner, some later. There are no sewers; the water is polluted. The infected wastes become fertilizer and then the veggies are washed with the polluted water. Ice cubes for drinks are frozen, polluted water. But twenty grueling minutes on a bus, weaving through blocks toward the central depot, offered no options, except a very clenched sphincter. At the depot, there were no bathrooms, but a nearby hotel offered relief.

We roamed around Tehuacán seeking lodging. On a limited budget, we only required hot water. After dinner, it was time to relax. The next day, before noon, our mechanical problems should be almost over.

A long day had increased the sleep time. I got to the VW dealer barely in time to catch it open. "No sir, we have only been open for a year. No, we do not stock any parts for a '65. Don't despair, they will have it in Cuidad

Puebla. It is the Detroit of Mexico. They have a Volkswagen factory there. It should reopen Monday morning."

A weekend in Tehuacán didn't raise my spirits, so a multi-level tour was in order. We found dull, yet, nice churches. Then, we checked out a foreign movie dubbed in Italian. We couldn't seem to win. On the square was a fantastic restaurant, and Saturday night, it was what was happening.

We were a disheveled couple, easily entertained by the antics of an anniversary party. A local couple honored their fiftieth in full plumage. After we'd dined, a local offered more drinks. For the sake of US foreign relations, we accepted. Enrico, the gracious, swing-belly host was a watch salesman who knew what was going down.

Broke down in Huautla, eh? As soon as we told people, they'd look at Sally's swollen nose and howl with laughter. No, of course we didn't get high there. While Enrico tried in vain to scoop Sally, I cordially drank for diplomacy. We were okay to the extent that Enrico's company picked up the entire dinner tab.

Sunday in Puebla was bland, with nothing to really explore. It was gray, like Pittsburgh.

Monday morning was spent at the VW dealers. Yes, plural, there were two and neither had the parts. Seems vans weren't imported until 1966 and mine was out-dated by one year. They were reassuring. They would call Mexico City. They would definitely have the parts. It would take two days just to compare part numbers in a warehouse.

The typical Mexican response was frustrating. Then, I circuited every used parts yard in the vicinity. Mexicans recycle everything. In such a low-income society, everything had value. Finally, I discovered one; the only one in the entire city, one used axel, for thirty-five dollars. Two drawbacks: I'd gotten the part numbers from the first distributer in Tehuacán and the used one had only one of the many serial numbers different. One number! What else? The bearing was bad. Another twenty bucks, and it would be ready—hopefully—to have the good old Volks rolling again. Then, the parts place wouldn't take traveler's checks!

Nerves frayed, I bought the axel and taxied to the bus station. The axel and bearing weighed about fifty pounds. Lugging that cumbersome weight around wasn't the easiest thing. Finally, a lot of ducks were again realigned and I headed for the bus to start the trek back to Huautla. I was told they only sold tickets on the bus, but they didn't and we had to wait three hours for the next bus at ten that night. This would be a night passage.

After a short night in Tehuacán, and on to Teotitlán, we reached the stop two hours early for the bus to Huautla. We had no diversions and were still in doubt about the parts. Another Mexican freak decided to be our friend and accompany us to the village of hongos. Between Sally and him, there should be no problem getting our points across.

Everything was as good as it gets on a bouncing, hot bus ride until an old man with dead eyes began to babble at me. I shrugged and ignored to no avail. He poked and pestered until I faced his putrid breath. I was his 'paisano' and 'Italiano'. For some reason unknown to me, he thought I had been coming to meet him. He continued harassing me until everyone sitting close by was watching us. Finally, the old man bent a bit and whispered into my ear, 'Heroin – Opio'. He thought I was his contact, traveling to buy his crop. From the look of his eyes, he'd sampled too much of his product. "Blue-eyed Italiano," he kept muttering and pointing at me. "Paisan, quince mil pesos por kilo."

I couldn't believe it, being offered a kilo of pure Mexican smack for a grand. No more conversations with this guy were going to benefit anyone. Smack has too much bad karma. The old loony thought I was holding out for a lower price and dropped to twelve hundred pesos. He refused to stop pestering me. Finally his stop came and he departed alone into the hillside. A man sitting beside me tapped and said his brother was a Federale and that the old man was considered a menace.

The bus made Huautla at sundown with the addition of a chilling rainstorm. It actually felt relieving to return to the weird village of the mushroom cult. Trudging the heavy auto parts up the muddy street in the rain wasn't much fun. The van was still there and Bill was asleep on the rear cushions.

Poor Bill had been deserted a few days previous. The local officials cordially asked the girls and Victor to leave because they had overstayed their permiso. The officials hadn't hassled Bill at all. It may have been sympathy for those less fortunate. Mike had decided to follow the women.

Bill was looking a bit forlorn. For two days, he'd been drinking coffee, reading old National Geographics, and talking with himself. He was freaked out by the Indians, who never permitted him any privacy. Bill was the Americano of the Huautla roadway zoo. In the days before easy personal communications, Bill had expected us back in a day. It had been four days and it had rained continually since we'd left. The village was all yellow mud. The following day, the van would be fixed; we'd visit María, and then depart for home.

HURRICANE BRENDA

We didn't know it, but Hurricane Brenda had made landfall in the Yucatán and was headed our way. Brenda was the first big storm on record to have ever crossed Mexico so far south. As the storm moved west and the eye passed south of Huautla, it brought the worst flooding the area had seen in twenty-five years. Not being fluent in Spanish, and with no television or radio, we didn't have a clue what was happening.

The rain continued the next morning, but this had to be the day of truth. Would the used parts work? By noon, I had merged the old with the new and realized there was one more bearing necessary. One part was all we needed to roll, but it was in Puebla. After days of hassling, rolling in the muddy road, and being quizzed by everyone, it was close, but not complete. Bill volunteered to do the traveling this time. I couldn't face the mountain bus ride again and Bill couldn't handle the local zoo effect.

A full day of street mechanics was exhausting especially during a torrential rain. I had no fire to dry myself. I changed clothes and with Sally headed to the Huautla Hilton, owned by El Presidente. The room had a hot shower and a great view of the mud puddle that had been the municipal square. I'd planned on Bill returning in two days, three at the most.

Lying in the mud, combined with the chilly temperatures, had tightened my back. I paid a visit to Mr. Gomez, the pharmacist. Two for a quarter was the cost of his best pain relievers. I washed down two and a vitamin C with some Kahlúa and didn't see any more of that gloomy day. Sally curled up as we'd clocked a lot of miles up and down those mountains during the previous week.

The next day, we slept until minutes to noon. It was still raining. Pneumonia hatches in weather like that. Everything was damp. The air was cold. There was nowhere new to explore. We'd seen everything except María. I waited for her return.

I called the Puebla VW Parts dealer and they hadn't checked for the parts yet after two days. The bearing place said they could send them COD. I decided to have them ship the part in case Bill didn't get it at the dealer.

The restaurant's service had declined a direct relationship with my tips. The owners knew me by name and at every meal; they'd talk about me with their other customers. I'd hear my name and look up to see many faces with gold-toothed grins. The street kids didn't even bother to hustle me for mushrooms. I'd become an accidental resident.

Tropical Storm Brenda seemed to stall over Huautla. The rain had wiped out the telephones. A voice from anywhere would have been cheerful. The village no longer seemed like the Garden of Eden. The road was too

ROAD TRIP: HUAUTLA

dangerous for the buses to run.

Sally's visa was about to expire. Our plan had been to give her a ride back to the Burgh to save her money. Luck intervened and she met a couple in the restaurant. The man was a pilot for Mexicana Airlines, visiting Huautla with his American girlfriend. They welcomed the company on the muddy road. I was sad to see Sally go. She had been the best translator and had become a close friend.

It was the third day since Bill had left in search of parts. Now, I knew the loneliness he'd felt when our return had been delayed. Huautla was a very strange place. The first impression of the real paradise dissolved into a dissolute feeling of weirdness, almost like the Twilight Zone. The drug-induced euphoria slipped away. The rain and gray days without any companionship led to depression and sadness. The local Indians were different. They never seemed to converse much. I got the feeling the Indians could communicate by telepathy. The local Mazatec spoken language was more like a chirp, in short, very accented sentences.

The rain had eased. I almost expected to see a dove with an olive branch at my window and Noah's Ark nestled below in the valley. With a bit of the sun peeking through, there were smiles at the restaurant during breakfast.

I mucked it up to the van to discover the torrents had not washed it over the road's edge. Everything was still secure. The creepiness of all the locals wouldn't permit me to extend trust. What we had in the VW was a treasure trove to them. The van was loaded with everything they needed: blankets, a cooking stove, implements, tools, and clothes. If our despair had occurred anywhere in a big city, our vehicle would have surely been sacked after being parked for three days.

In a day or two, we'd either be driving out of Huautla or riding the bus and abandoning the van I'd grown to love. Everything needed to be sorted, repacked and operational. I was optimistic that Bill would get the correct bearing. The reorganization was something to keep me occupied. It felt as though I was making amends. The time seemed right to make an offering to the gods I may have offended. Anything that wasn't necessary became a gift to anyone who would take it. The Indians were gracious and loved the glossy National Geographics. Their children formed a line and were given combs, brushes, clothes, whatever needed a new home. The two families where the van rested didn't understand what I was doing, but they accepted the generosity. It was a holiday for the neighboring families. The crazy gringos were having a giveaway!

Once everything had been reorganized, I walked farther up the street to snap some photos, fearing another downpour might erase the chance. About another mile up the hill, I still didn't have a clear view. The road curved and the view demonstrated the sprawl of this hidden mountain village. Two young locals approached from up the road. One, obviously drunk, demanded I stop taking pictures and tried to grab my camera. Pushing him away seemed to throw me into yet another Mexican mountain extortion scheme. This fellow was almost a duplicate of the first stubby, black-Banlon type who'd hassled us when we were tripping. The drunk's babble easily translated into 'hooray for me and screw you gringo'.

Huautla Give-Away Group.

The other man tried to quiet the confrontation, but the drunk kept forcing the issue. I owed him money. He demanded I pay him four dollars or accompany him to El Presidente. As I turned to walk away, he grabbed my arm and flashed his official sanitation department identification. This was supposed to impress me. His younger friend finally separated us and they moved down the hill.

Not wishing to follow those two down, my climb continued up the street. Two bends later, the younger man caught up with me. He was the true con man. It had been his plan to be the good guy, get the drunk to cease irritating me, and then return for his reward. His approach had to be respected, as cultivated and non-threatening. His fee was a reasonable five pesos. Following this tact, he could easily become a future El Presidente.

The view was well worth the muddy hike to the crest. In all directions were ethereal mountains and lush valleys all crowned by fluffy clouds. This could have been the top of the world. That moment was exceptional, realizing if I went any further into the mountains, I would be coming out the other side. This was the crossroads.

The road from the village below met another track that ran along the ridge. There were only two houses at the intersection. Residents of both homes were eyeing me. The one to my left sent an envoy.

"You came to meet María?" he asked.

"Sure." I'd had no idea where María lived, but my route had led me to her door. That summed up our entire journey. The scene was extremely strange; the couple and the other house didn't feel real. There was no conversation, but my intuition buzzed me they were characters. An Indian man

ROAD TRIP: HUAUTLA

and woman in their late twenties seemed to be an advertisement. He was in spotless, white farmer's attire and her dress was finely embroidered. They were showing affection by holding hands in public. Their home was a mirror image of the one I was being led to. Both homes were impressively big, maybe twice the size of the homes on the lower levels of Huautla. Each was brown adobe and had an expensive corrugated steel roof. All others, except businesses, were thatched. Suddenly, it dawned on me; this was a business.

The Indian explained he was María's son, as was the man of the other couple. He went on to relate that María was very old and everyone on the mountain slopes, and in the valleys were her relatives. I did not realize it until this research, but he was probably saying that was where her children, the hongos—'ninos santos' (holy children)—were grown.

Faced with meeting the boss sacerdosita alone, I was a bit frightened. I needed to reevaluate and summon some courage. Everything that could possibly hinder my mental faculties had been ruled out. It was time for the extra-perception ride of a life time. Huautla was probably a once-in-a-lifetime journey. It was the conclusion. With or without the van, we would be departing. It was time to indulge.

I couldn't help but wonder what María would be like as I walked to her home. One thing was certain: she was mystical. That was her fame, all natural and the oldest, most trained of the sacred hongos women. Traveling for parts, I'd mentioned being broke down in Huautla; almost everyone knew of her. Each claimed she was the 'highest' in the western world. The house on the mountain crest and the view inspired the moment.

Her son asked if I understood about the hongos. I nodded and was delivered to the front door of the mud brick house. He left to get María. He did not enter the adobe, but went around the left corner to an almost hidden, adjacent wooden shack. The wood building was older with wide gaps between the black, weathered boards. The small house's roof was thatched, and burlap rags draped over the windows and doors. Her son explained that María never had felt comfortable in the new house.

He called something in Indian twice. An old Indian woman came into the daylight. I realized I had seen her, before the rains carrying a load of firewood with one of those forehead straps. That had been miles below, at the municipal square. María was weathered, and looked to be in her sixties.

Too many years later, while writing this book, Mr. Google informed me of many previously unknown facts. María was born in 1894 and made her final crossing on November 1985. She was a healthy 79 when she tripped me. This diminutive woman, literally in the middle of nowhere, combined with an unlikely gent, Gordon Wasson, in 1955 and had propelled our entire world into a new era—the extra-perception of psychedelics. Our universe would never again be the same. (See LIFE Magazine, May 13, 1957.)

María didn't appear much different from other village women. The first sacred woman who had sold the hongos to Sally and I had looked more affluent. The head sacerdosita appeared ancient and indifferent to all things. She extended her wrinkled, brown hand to me. Touching her, I caught a jolt to my nervous system. Her incredibly soft hand filled me with a tender, flowing feeling—a body rush of pure energy. Sentimental is the only descriptive word that seems fitting. It was as if I had known this old woman all my life. Her eyes knew no questions. I was certain she knew my story. The five-foot tall woman was my friend. Her gaze relieved me of any worries or cares. My spirit swelled because I'd just met one of the world's most knowledgeable people.

I'd encountered politicians, artists, performers, and thinkers in the States—people who had been told they were great and should be treated with respect. That flashed across my mind for an instant with their clothes, jewelry, unnecessary possessions, and their insecurities. María was the opposite and needed nothing to be secure. She was distinctly one with the forces of nature. Easily, with the right promotions, her fame could have equaled that of the Hindu gurus, who were then hawking soul-relieving meditation. Her existence could have been better, but she'd chosen to stay just as it always was. The deep, brown-eyed stare signaled she was not concerned. She was both of time and timeless. The past and future had no effect. For me, this was the time.

María Sabina

One photo was permitted, something the other sacred women had prohibited. Her son squeezed five pesos for the shot. Everyone in Huautla got beside María to pick up the money she attracted. Her family was no different. María stood against the adobe wall in her soiled ankle-length dress as I snapped.

Her son led me inside the adobe house. It didn't seem lived-in at all. It was the station where the 'Hongos Express' was about to leave from, or it was also the 'hongos clinic'. Two wooden frame beds, a rickety table with two drawers, and an old phonograph with a loud speaker were the only furnishings. That was a lot for an Indian household. It was all very neat and clean. I sat on one of the beds as if awaiting a doctor.

María entered with a parcel wrapped in newspaper, definitely the medicine. She handed her son the packet.

He unveiled the 'shrooms and handed them to me.

"Hongos, señor, four dollars."

This was a bargain. María was giving me the works for a third the reported fee, or maybe it was a rainy-day sale. I eagerly consumed the tasteless, dried mushrooms.

The priestess chirped a few syllables to her son. It was he who told me she was not permitted to speak to me or to speak Spanish. María kept to the 100% Mazatec ritual as her only language.

"María wants to know what took you so long? You have been here for more than a week."

The old woman knew all that transpired in Huautla. This was her domain.

"Where is the girl? Are you married?"

Beginning to feel the hongos, I mumbled an explanation. María sat rigidly on the other bed with her spindly arms crossed ladylike on her lap. She seemed to understand our conversation, yet her son repeated my answers in Mazatec.

When I finished my tale of the journey to Huautla, María again asked, "What took you so long to come?"

I had no answer and shrugged.

María stayed with us for a short while, then rose and left. Her son explained she must eat and freshen to assist me on my journey. He left and I looked for something to hold my elevating attention.

The plain room held no clues to hongos, religion, or mysticism. With the hongos edge, the view was even more amazing. It was about one in the afternoon and the sun's heat had dissolved most of the mist. The weather had changed for the better, but it was still gray. The room was about twenty-by-forty and empty of belongings except in the corner next to the chest of drawers. The beds were entirely constructed of wood planks, with foot-high headboards without any carvings or trimmings. Next to the beds were two wooden benches holding stacks of adobe bricks.

My central nervous system was beginning hongos calisthenics. My mind was searching.

An embroidered white cloth was draped over the table. A bamboo rod leaned against the wall partially obscuring posters of two unfamiliar people. The creepy case with a peaked roof held a wooden crucifix with a golden Christ. I'd never considered María to be the one who'd embrace Christianity. On top of the case was a lantern.

María's Hongos House

At the other end of the room was a porch. The silver sun was heating the valley, burning off the week's rain and misting the greens into shadows. Some of the wet green shined, but the constant breeze made the scene seem to vibrate. It was the first time I'd actually watched clouds form. The mists floated up, whirling until a cloud bloomed at the ridge.

The edge of the porch was the edge of my world. The ground had slid with the rubble piled about a hundred meters below. The house sat on the precipice, with no backdrop. The blood was pounding in my head. I returned inside and sat on the bed. A black-and-white dog, same as the original Spot, in the Dick and Jane books, came into the room through the middle door and trotted up to me. It was affectionate just as most of the dogs in Mexico had been.

Alone in the room, the pup was a distraction. I petted it and scratched behind its ears, rolling him over to scratch his belly. Spot scampered around, happy with the affection and disappeared beneath the bed. He was under there for a bit and my drugged attention floated away. Spot was forgotten.

A small, black-and-white pig appeared from under the other bed. My mind was murky and I remembered the dog, but I couldn't discern if the black-and-white hide was similar. The pig snorted and approached with short, stubby steps. I'd been sitting for so long that when I tried to move my arm, I had to make a definite effort. My arm just couldn't raise itself. My conscious mind had to instruct the muscles to lift, stretch, and grab. I was feeling pretty crocked.

My hands ran over the back of the pig, ruffling the coarse hairs. The pig loved the attention, not flinching. My fingers reveled in a refined sense of touch. I could feel my own skin. The pig left. I must have spaced out and was enjoying the sensations from just sitting there, sort of checking in with various body parts, feeling this and that.

I don't know if I consciously reclined on the bed—or if the weight from the hongos had pushed me back. The wavering grid of the wood rafters held my attention. I felt as if my physical systems had been shut off. I was completely relaxed—while only a slight portion of my inner mind remained awake. My eyes were working, though. The more my vision focused, the more I comprehended the confinement of my body.

Something bumped into my legs and the feeling returned. Inertia seemed to work on my spirit. As soon as a spark of energy touched me, I became active. I sat up, absolutely spaced out. The black-and-white dog was back, darting and sniffing. Without any apparent lapse in time, the dog ran under the bed—but the pig came out the other side. It pranced over, circled my feet, and modeled the change. The oinker did the presto-chango—again shielded by the chest of drawers. The pooch reappeared. I was thinking about how neat this hallucination was. This was the big time.

The dog hung around for a short while sniffing in the far corner, and exited the middle door. I didn't feel as high as when I'd been lying on the bed, probably becoming accustomed to riding the hongos wave.

Two small children, a boy and a girl, popped into the doorway. They sort of shuffled around. The boy had his hands in his pockets. They approached, pointed and said, 'Hongos'. I shook my head and murmured, 'María'. This was the first time I'd spoken and it felt like a learning process, something new to me. My Spanish rolled easier; my comprehension was enhanced.

The kids fingered my camera, which I'd managed to keep on my arm. They said they were María's 'bisnietos' (great grandchildren). They were willing to pose for a peso. The kids were cool, photogenic, and I was feeling like the last of the big-time spenders. They reminded me of Dondi in the Sunday comics, like raggedy orphans. They played, dancing around and toying with the coins.

I stood, staggered, and leaned against the middle doorway. Outside, an intense game of marbles ensued among the family. There seemed to be so many members of María's family, they could have had a regulation football game and still have enough for a cheering section. The most accurate marble shooter was the five-peso, two-stage hustler I'd met lower down the road. He was another of María's grandsons. He'd gotten sauced with my money and was now my long-lost amigo. Mr. Two-Stage saw my glued on smile and offered to display his famous grandmother's mementos. He opened the chest and exhibited a poster of Morrison and the Doors. There were photos of María and people he couldn't fully explain. I tried to focus. He withdrew an album of María's chants, which I had not yet heard.

María's Great Grand Children

Next, a mature Indian couple approached. They wanted to sell marijuana. The woman slid by me into the house and beckoned for me to examine her goods. The man came up behind me; I felt sandwiched. Their weed was wrapped in paper and the smell was intoxicating. My perceptions were more than slightly impaired. It flashed through my mind, this could be another con. The man kept tugging at me, much too anxious for the purchase.

"You like the mota, señor? This is primo mota. It is our own. We grow it here." The Indian claimed his wife was also a sacerdosita and her weed was the best.

"Quantos pesos?" I asked.

The team wanted five dollars (sixty pesos) for the clump. That was a hell of a lot compared to what we'd gotten before. I declined the offer to roll and smoke one. I didn't want to get any more confused. It was all too hectic.

They exited, but the woman returned. She extolled the quality of her herb and we agreed on an acceptable two dollars.

With the smoke deal, the drunken grandson, the kids, and the dog transforming into a pig, my mind was reeling. I was beginning to lose my grip. A dark evening was moving up the valley. This gauged my stay at about five hours.

Everyone had vanished and I was relieved to be alone. Again, I felt the locals feared the darkness, or something that came with it. During our stay in Huautla, we noticed no one was outside at night. It could have been the primitive surroundings or just their primitive minds, but as light disappeared, so did the people. My thoughts expanded until I felt my created evil approaching. The room was dark and my eyes only saw a speckled pattern. My spirit felt as if it were sliding into melancholy. The night's fog was deflating my high.

María returned carrying a bundle of clothes and indicated I should change. The shirt and pants were what the Indians wore, soft and loose. They draped my body. The suit felt good, like my body craved it. The clothes were clean and plain, void of embroidery. She stretched a woven straw mat on the dirt floor and handed me another packet of hongos. These were fresh—but still not the shape of the ones Sally and I had ingested. I could feel my consciousness accelerate.

The high priestess puttered, readying the necessary items as she must have done a thousand times before. María lit a candle and came over to sit by me. I never saw her eat any mushrooms. In retrospect, by that age, and after so many trips, she could probably transcend without any assistance. She felt my arms, running her hands along the soft underside. Somehow, by her touch, she could sense the elevation of my condition. A short while later, she brought another dose. These were the tiny black-and-white penis types. This clump was horribly acidic or perhaps my taste was more sensitive. I had trouble swallowing, couldn't make my throat work, and my concentration was fleeting. María led me to the mat, getting me to kneel.

There was nothing except the candle.

María knelt facing me and extended my arms to my knees. This position locked my elbows, keeping me from falling. She pushed up the sleeves of the shirt and applied a salve to my lower arms. It wasn't menthol, but it caused my entire body to warm. These were the last memorable body sensations. The candle flickered. María began to chant. I remember thinking it seemed like a vibration feedback. She continued rubbing and chanting, both were in rhythm.

The shadows dissolved. María was gone, yet the chant remained. I was at ease; my mind was receptive. Whatever I was, my form was gone, shed; I was pulsating. Like a caterpillar, I gained new wings. It was good, everything was good. My consciousness was rising and then sliding down invisible slopes. Like driving a car where your stomach feels queasy at the bottom of a quick hill and then you continue up. If you gas it, the queasiness increases, but if you slow, you miss it. I decided to step on the gas and get the most out of this hongos ride.

The chant was now in the daylight. The chant was only directional, without form. My view was from above, overhead. I focused and choices were made. Any situation or location could be experienced. I remember shooting down into a damp forest, darkened by the leaves overhead. The trees grew to seventy-feet tall before they became a ceiling. I floated between the thick trunks, a meter above stunning mossy rocks. I didn't hear the chant and hoped I wasn't lost. I realized my spirit and body were in different places. A twinge—it was fear. Was I lost?

I sensed I wasn't alone. My mind perceived the location as entirely friendly, maybe beneficial. Getting a feel for my new being, I explored a bit, experimenting to see if I had other senses in this vision. The methods that my earthly body used to feed my mind emerged. This was the freshest air, and cool due to the slight breeze generated by my movements. Sweet vapors registered.

Instantly wanting to sense the rocks and moss, I did. Not by extending an appendage, more like by being it, feeling the essence of the moss. Through the moss I intuitively sensed the black soil gripping, holding everything. Nurturing. And the rocks, solid earth, the strength of solid, compressed energy. They were as nothing else had ever been to me. I felt time in the heart of the rock, but more, complacency—as if no more transitions would be happening. No more new startling inputs, just truth.

Something communicated with me, building slowly. First, it was in whispers, demonstrating, and teaching. Then I heard the words form, or rather, the thoughts. It was a new association—there, but not visible. It suddenly came to me, I was not visible. I was a beam.

The other thought finally registered, but it seemed as though we'd been exchanging for some time.

"This is nature, builder of worlds," the melodic sound emitted. "The trees, soil, and rocks and moisture aligned, everything smoothly existing in balance. The tree is a combination of all, with its radiant spirit. The tree dissolves, held in place by the rocks. The soil eventually has its growth, energy, and years packed into it—until it becomes the knowledgeable, invulnerable rock, unmoving and only vulnerable to the spirit. This is pure, naturally rising into one of the four changes; the seasons of the living world unveiled, the germ, growth, experience, and rest. The four are everywhere to be witnessed. Every growth is in a stage until it finally attains the rest of the rock."

"The restful earth has no weakness to the other elements: climate, water, fire. Any conscious growth is unsafe from these tools of the spirit, the master. Fire destroys—rampant energy sucking the glow from life for an instant burst. And liquid—feeding and helping until over-confined, generating tremendous energy. These two, coupled with climate, when unleashed, breathe greater energy. You recognize it, don't you—the cycles?"

"Your self breeds your educating growth. The soil in which you are planted, its richness of spirit, develops each positive or negative. You self is either blossoming, happy with well-being, or wilting. Everything is inside and there is no bottom or sides to the inside. Inside is forever."

With those words, the landscape became the white sand edge of a deep blue sea. I sensed warmth. "How does this happen? What do I see? The changes?"

The answer rolled, "Everywhere, everything, all things are inside your mind, all experiences of your past and distant past. Your essence is from forever. We are tapping that information."

"Everyone who asks the questions of the how and why of life, are, and have been since before the change. Within, from some point of expansion, lies the answer to every possible doubt. Each is planted with the purest of essence, knowing, having just resided with the spirit. Our glows are warmed and beautiful with contentment as we watch our blossoms and bloom. Seeing so many. Don't doubt, just accept. We diffuse as the sea into the rain and into rivers, back into the sea. The spirit is the sea. We wander about, surfacing among tricksters and charlatans. All is a maze leading to the negative route. Our goal is to avoid the negative, the sickness, to locate the spirit's goal. All the while, we build our fire until the energy is enough to be attracted back to the spirit. Everything is energy. We are energy. All logs in a fire burning alone and together, burning bright."

"Our mind projects the puzzle on other energies. Whatever we create is a second generation. The inner feelings—disturbing, distracting. Yet others are ready to receive. Most are undiscerning of the negatives; few emit the truth. As each conscious world evolves, the more paths there are to choose among. None leading to the absolute positive. Difficult to remove the doubt, passing with a dim, unfulfilled glow."

"Ages bring cycles to every world and each circulates back unless harmony—unison—with all is achieved. Not so much the importance of the single being, but its enhancement through and with other energies. More logs on the fire burning together. Signaling attention. Understanding. This moment is exactly that. Our fires are merged together and we read each other's histories. Letting the wind of knowledge show our affection in the sea. If five were joined, it would be magnified."

I listened. Each thought generated questions.

"Who are you? I cannot discern features. I need to view you." I asked.

"See me how you choose. I may appear in any form, but the spirit is invisible. I am you, I speak your words. You are comfortable."

The voice continued, 'You met me as an old woman. You respected me. I tested you while in the shape of animals and of children. Both appearances you treated with respect and kindness. My forces described your being basically positive. Those experiences were the soil that nurtured these. If you had been of other shades, I would have demonstrated the facts in more frightening ways. However, you are very receptive, but still not assured I am not negative. You will carry that thought as your life's question; is what I speak true...or created to confuse?"

Goats appeared foraging on a sloping pasture. I could taste the chewed grass and feel myself slide on hoofed feet. I noticed the change and was impaired by the creature's shell. A sense of danger poured over me. A great bear was crossing the field. I was uncertain if I was really there, but decided to flee, hopping on spindly legs. I was overcome with doubt. I just ran until I'd been backed up against a rocky crag. I was cornered by the bear with only the alternative of leaping into space.

Instantly, I was soaring again, but felt isolated, and unattached.

"You see, we can choose any shape; that experience was necessary. Too many put themselves on those rocks until they must leap. Hopelessly clinging to false beliefs for protection. If you believed the bear was harmless, you would have coexisted and become educated in the ways of the bear. Instead, you ran in fear. The other creature only followed, perhaps expecting you to lead it to its next destination. In the end, you panicked, cornered; you attempted no communication. The puzzles are real. Each must be reasoned before we conclude we are trapped. Traps are plentiful and you have run into many in your consciousness."

It felt right, everything felt right.

"Observation will show the right direction. No hurry—watch and learn. Make the correct choice. When you hurry, the flame flickers; slower, the flame burns bright and steady."

We were rising, drifting among the stars, towards one radiating source of white light.

"This is the flower of the master's garden."

"A star?" I asked.

"Look closely at what it is made of, all the reflecting particles of the whole. Examine the edge."

My attention focused on one radiant. The more I focused, the white became colors, a rainbow glow. As I approached, the colors became balls suspended in rows. The balls weren't round, but of every geometric shape. This was what I had envisioned during the other hongos voyage.

My guiding narration returned. "This is the goal: to be part of, as light in the night sky. Brighten the darkness. Radiate. Notice how each glow locks into the final form. Harmony without friction. Nothing except the mover can act upon these. They are complete."

The shapes separated as we dove into them. Surrounded by brilliant oranges and greens, I sensed higher energy at the core. Attracted. Drawn. A bright orange triangle was ahead. As I plunged into it, I felt depth. Some type of pressure. My energy tasted it, merged as if it were water, swirling and bathing. I sunk until the orange vanished, replaced by black darkness.

I awakened on the straw mat covered with old blankets. It must have been late at night. Everything was black and I was dazed. I'd hoped it was night, and that I hadn't gotten stuck in some part of another perception or dimension. María was gone. Her job was finished and she was probably sleeping. Nothing was moving, no noise. I pulled things together and tried to stand. Lightheaded, I banged into one of the beds and sat down or collapsed. I was breathing hard. Even though I was awake, I was damn high—seeing funny things high. The more I grew accustomed to my body, the more I felt the hongos. Hissing sounds from the darkness echoed.

The peasant clothes were soaked with sweat. The night should have been chilly, but I was too stoned to feel it. Paranoia began. I felt loneliness, empty, abandoned as my spirit cooled further from the trip zone. Changing clothes wasn't easy. Every piece felt coarse and smelled. I guessed it was my smell. My fingers ran over the jeans, my T-shirt, and neither felt comfortable. The more I stayed within María's house, the less I wanted to.

The pajamas came off and I used them as a towel. In this condition, how would I leave? Could I leave? Focused, I'd managed to keep my camera and reasoned I knew the way out, down, down the hill, down the road, but I wanted to be down, down-down.

Slipping into my boots, I heard talking from outside. Three shapes appeared in the doorway. Three of María's grandson's had returned and offered to guide me down to El Centro and the hotel. That sounded like it had all been planned, included in the hongos experience. The brothers got me to the dirt front yard and took a new path.

Once nothing was in sight, they hit me with, "Five pesos, señor, to guide you to the road, or twenty to the hotel."

The bastards had duped my stoned ass and led me into a maze.

"Okay, five pesos to get me in the right direction."

The older two left the younger to the duty. He led me over a grassy embankment, down a slope, and deposited me onto the main path.

"Bear to the right señor," was the last advice I'd get from María's family. Frightening and nerve-wracking described that moment. I was alone in the night on the road above Huautla. With every step, I imagined movements on all sides—creatures or worse, ready to pounce.

Trudging, staggering, I found my Volks. The familiar metal brought on shuddering thoughts of the pending difficulties. Stuffing my hands into my pockets, I discovered the newspaper-wrapped pot. Obviously stoned, wandering the streets of Huautla, it seemed best not to be holding marijuana. I stashed the weed under the van's front seat and plodded down the hill. The few street lights brought me finally to the sanctuary of the hotel.

Not wanting to draw any attention, I carefully felt my way up the steps, tip toeing and running my fingers along the dark wall. The mayor and his family didn't need to be rudely awakened by something being knocked over. At last, my fingers found my room's door. The door creaked revealing Bill's return. Not alone, though, as two more shapes were in my bed.

The little light from outside showed Bill's shape and I shook him awake.

"Where've you been? Got back about eight last night and you were gone. I got the bearing, so we should be out of here in the morning. You been crashed in the van? How've you been occupying yourself?" Bill asked.

His watch read two in the morning, as good a time as any to chat. Still reeling from the 'shrooms, I was at a loss trying to describe the María effect.

"María sounds fantastic, but I don't know if I could handle it." Bill continued, "You'll never believe what happened. The bus was late getting into Teotitlán from Tehuacán so I missed the one back to here. This couple and me (the two shapes in bed) hitched a ride in the back of a pickup truck. You can't believe how fast these Mexican cowboys take that winding road. My back will never be the same. Anyway, about an hour into the mountains, we spotted a red light in the sky. The light was hovering in one of the valleys, sort of pulsing. We watched it for a while and then it must have noticed us, maybe the pickup's headlights. The red light closed in, following. Scared the hell out of me. First flying saucer I've ever seen."

"The truck driver went nuts when he saw it—plowing so fast, we had to lie down in the bed to keep from bouncing out. I'm sure he set a record." Bill was pumped up. "The ride took only three hours to get back here. The man sacked out in the bed told me they see those lights around here all the time. They are especially common now, in the rainy season."

Flying saucers and magic mushrooms seemed to go together. I was still in the 'shroom zone. The wheel bearing meant we'd be leaving these mountains, but I was certain they wouldn't be leaving us. I could hear the words, "You will carry that thought as your life's question; is what I speak true…or created to confuse?"

* * *

Morning came early and we tromped to the van. Within an hour, the job was completed—but not perfect. The Volks would have to limp the distance since only three of the wheels would have brakes. The line at the bad bearing wheel had broken so it was crimped with Vise-Grips for the entire return trip to the Burgh.

Bill, listening to a Little Rock, Arkansas, radio station for company while he was alone, had run the battery down. Pesos enticed helping hands from our Indian neighbors, and on a roll, the van started. After an over-extended stay in Huautla, the mushroom village, we were finally on the move again. One petrol stop and we were headed down the road and out of the mountains.

At a tight bend, a local hailed us from the roadside. From under his shirt, he produced a large psilocybin mushroom. "Hongos grow big during the rains," the farmer explained. "You can have this one for fifty pesos."

Neither Bill nor I had any immediate used for more 'shrooms. Declining, we drove on. At the next bend a group of young girls waved us over. They led us to their mother's isolated house. Two older women were stewing dinner while trying to up their income. We had to again refuse mushrooms being low on bucks. These were the potent clumps—and after some bickering and excuses, we made a deal. The old lady said we could preserve the hongos indefinitely in honey, retaining their potency. She also sold honey. The deal was five National Geographics and two dollars. The next problem was, where to stash them. We had to bring some souvenirs of Huautla back to our stateside amigos.

Our route took us to Tehuacán at dark, without any more hucksters or mishaps. The brakes worked fine.

The following day, we cruised Puebla for more trade goods to cover our added vehicle expenditures. We found some beautifully woven blankets and we headed to Guadalajara, hopeful for more exotic items.

Somehow, I remembered the Huautla pot I'd purchased from the woman at María's. Parked on the outskirts of perhaps the most beautiful Mexican city, we tested the weed's potency. We hadn't smoked half the joint before we'd been enveloped in a pattern—paisleys again—and soon passed out.

The next day began with incredible news on the radio. A sizable earthquake had struck Puebla, where we'd been the previous day. The quake had been quite severe, leveling several buildings and striking another city, Orizaba, on the other side of the mountains that contained Huautla. The rains of Hurricane Brenda had soaked the ground, causing landslides before we'd left. The quake could have proved disastrous for the hidden mushroom village. I envisioned a sorry fate for María's cliff-side adobe. We listened to damage reports through the day, but the obscure village was never mentioned.

The VW made it back to Pittsburgh and our 'souvenirs' put serious grins on the faces of a few friends. We related our tale a couple of times to mixed responses. Few chose to believe a mushroom-cult/religion existed, until my slide film was processed. One extra photo was with the roll. It was a pattern amazingly similar to the Mazatec embroidery stitch. There's absolutely no way I could have taken the picture.

Mysterious Pattern Photo

'You will carry that thought as your life's question; is what I speak true...
or created to confuse?'

MARÍA SABINA

I didn't know anything about María, except what our acquaintance Victor had told us in Huautla. Prior to meeting Ernesto and Pepe in Tampico, we'd never heard of the mushroom cult, Oaxaca, Huautla, Mazatecs, or hongos. Perhaps we were inexperienced hippies, not sophisticated in the lore of natural drugs beyond weed. We weren't virgins to the scene, just had never traveled in search of a new high.

When we returned home, The University of Pittsburgh's library had zero information on María. Since we had never heard of Gordon Wasson, there was no way of researching from that angle either. This was long before Mr. Google and the Internet. As this memoir progressed, it is amazing how María Sabina contributed to evolving our culture. To live so basic and so isolated, and yet impact the entire world, is remarkable.

María Sabina was born in Huautla to a long line of mushroom people. Her birth year varies, but most accounts say it was 1894. It must have been a rugged life in the mountains. Huautla is also known as the 'Eagle's Nest', and the location of María's homes with the incredible view must be its perch. This was most likely the property of her mother's family.

María's parents were subsistence farmers. Her father, Cristanto Feliciano, died when she was three. Even though his father and grandfather were healers, his illness must have been too virulent. Her mother, María Concepción, took María and her younger sister to live with her parents in Huautla. Opposing the stories we heard of María being flown to Mexico City to preform mushroom rituals, it is more plausible she never went anywhere that she didn't walk. When she was seven, her father's family inducted her into the mushroom curative culture. María became an expert at diagnosing and treating problems of the mind and body. When Gordon Wasson approached in 1955, María had already been practicing for fifty years! She was a Catholic and had successfully combined various elements of the magic mushrooms with European religion to please everyone. In turn, the local church did not consider the mushroom rituals heretical or hedonistic.

María was taught to use the mushroom practice for healing. From my experiences, I believe the sacred women are trained to look inside the people who travel to them. Mushrooms were used to treat physical illnesses. They were also used to treat psychological problems, but it is doubtful the tribe comprehended the difference. Every sickness came from inside the person. The mushrooms and the training permitted the sacerdositas to peer into the spirit of their patient. The cure supposedly comes from the other side—the spirit world—and is invisible except to those who are trained to see.

The sacred Mazatecs studied how to cure the soul, or spirit. Theirs were probably the first and only programs of drug-induced ritual psychotherapy—both single and group—where both the subject and the doctor were under the drug's influence. The mushrooms bridge the psychic span and permit the sacerdosita to examine the patient's physical and mental condition from the inside, while also giving the patient a voice to explain their problems.

This shy woman was the first to perform the mushroom rituals in the presence of white, non-Mexican outsiders. In 1955, María Sabina had a most-unlikely visitor. That man would change her world and also start the entire planet into a different cultural spin. George Wasson had a penchant for mushrooms. From various sources, I learned he and his wife, a Russian, started their mushroom fetish in the mountains of upstate New York in 1927. The Wassons became so entranced by 'shroom power, they traveled the world in search of information. Each region/country relegates importance to fungi. They penned a book, Mushrooms, Russia and History.

On the other side of the Wasson coin, he was a New York banker and had made vice president at J.P. Morgan during WWII. Still traveling for fungi across the globe, Wasson theorized ancient man worshipped the mushrooms that got people high. This was a huge change from the common train of thought concerning primitive peoples. According to Mr. Wasson, in Life Magazine, May 13th 1957 Seeking the Magic Mushroom:

Our surmise turned out not to be farfetched. We learned that in Siberia there are six primitive peoples--so primitive that anthropologists regard them as precious museum pieces for cultural study--who use an hallucinogenic mushroom in their shamanistic rites. We found that the Dyaks of Borneo and the Mount Hagen natives of New Guinea also have recourse to similar mushrooms. In China and Japan we came upon an ancient tradition of a divine mushroom of immortality, and in India, according to one school, the Buddha at his last supper ate a dish of mushrooms and was forthwith translated to nirvana.

'Shrooms have been revered for a long, long time and usually portrayed as religious—a pathway to meet God. Psilocybin could possibly have been the very first drug. Ancient people, bipeds, were constantly looking for food and consequently, ate some fungus. It wasn't accidental, as they ate everything that didn't eat them. From Wasson's research, it seems get-high fungus grows everywhere in the world.

Psilocybin is quite an experience for modern people who are prepared and awaiting the effect. Today, people proceed with an organized and expensive experience, most expecting a vision of God. Imagine a basic, unknowing biped with no conception of God; drug-induced euphoria would be quite an experience for basic creatures. Our ancestors would sit and eat until their bellies were full. Even if the 'shrooms were low quality, the quantity would definitely push some primitive Homo sapiens' buttons. Did mushrooms give birth to the first conception of God?

The Mazatec Indian tribe of the southern Sierra Madres in Oaxaca considered mushrooms medicine for curing illnesses. They only ate the mushrooms when something was not quite right—physically, mentally, or morally. Unless there was a problem, they did not take the 'shrooms, and never for pleasure or recreation. In such an isolated area, well into the mid-twentieth century, the mushrooms were their only recourse to medicine. The 'little saints', as María termed them, were not for fun.

On the trail of mushroom religious magic, Wasson was the first outsider to witness the Mazatec mushroom rituals in Huautla in 1955. The foreigner made it to Huautla, walked to the municipal building, and boldly asked where he could learn about the magic mushroom. An official (prying the lid off the foreign Pandora Box) showed Wasson where the fungi grew and then delivered him to María's door.

María accepted the introduction by the local official, who was her friend. The problem with the two outsiders was simply that they didn't have any troubles that required María's curative services. Wasson and his photographer weren't only the first foreigners to partake in the mushroom ritual, but they were the first that came just for the high.

Instead of the sacerdositas we met, in 1955 the sacred women were termed curandera. What I experienced with the chanting ritual guidance is named a velada. María introduced the first white outsiders to the magic mushrooms, straight-laced Gordon Wasson and his photographer, Allen Richardson, on July 29&30, 1955. For the story, Life magazine paid Wasson the tidy sum of $8,500, and Richardson $2,500. Great public relations had planned a lengthy interview about the story on Person to Person, a popular CBS television news program. A New York banker went into the mountains of Mexico. There, he ate mushrooms that were very powerful drugs and greatly altered his perceptions. He did this in a dugout basement with a sixty-one-year-old Indian woman, who would not speak Spanish, only the local Mazatec dialect. This was not only accepted by the 1957 audience, but considered high adventure.

Wasson became a cultural superstar because of María. Aldus Huxley and Timothy Leary held his story sacred. Leary went to Cuernavaca in 1960 and ate the 'shrooms without guidance. He returned to Harvard, and with Richard Alpert, began the Harvard Psilocybin Project. Both were dismissed from the school.

María permitted Wasson to take her photograph, only if he did not reveal her name or the name and location of Huautla. Reading the Life article, Wasson had been tripped by Eva Mendez. There is no mention of María Sabina. Huautla is referred to as Sierra Mixteca.

According to Feinberg's Devil's Book on Culture:History, Mushrooms,and Caves in Southern Mexico, within weeks of the publication of the Life article, a San Francisco photographer, Howard Taylor, travelled to Oaxaca seeking Eva Mendez. Somehow, Taylor saw a photo of woman wearing a traditional dress from Huautla. Each tribe's embroidery stitch is unique. Taylor dealt the historical blow to Huautla's secrecy by identifying the location of the tribe who sewed that fateful stitch. The hongos were out of the bag and Shangri-La Huautla began to slowly disappear.

Hippies, like us, chose Huautla as the mushroom mecca as early as 1962. Tourism was the obvious profit mode for the Mazatec hongos participants. A rental house industry started. The attraction of many foreigners also attracted the Federales—who were mistakenly convinced María was a drug dealer. The influx of strangers initially confused the villagers. Of course, there was the lure of profit, yet the visiting 'hippies,' or 'jipis' didn't spend much money. More so, the touristas upset the natural balance of the Mazatecs and their customs were jeopardized.

María was hospitable to the outsiders to a fault. She remarked, 'Before Wasson, nobody took 'the children' simply to find God. They were always taken to cure the sick'. But the siege of the strangers continued. The village blamed María for publicizing the ritual and she was ostracized. Her house was burned (one account has it as her grocery store). In another account her son was murdered. After the magic mushroom box had been opened and attracted tourists, María Sabina lamented having revealed the mushroom rituals. The itinerant Wasson rationalized the avalanche of tourism-for-the-wrong-reasons as a mere contribution to human knowledge. María's reputation grew internationally, but diminished in her own village. Huautla became a tourist attraction, yet the tourists became a local distraction.

María, possibly, was not the highest of all the sacerdositas. On the wall of the mushroom store, we'd seen a photograph of a man who was a mushroom guide. No one in the village would give any information concerning who he was. It is quite possible that María's acceptance of Wasson offended those higher in 'shroom status. As Mexico and Indian cultures are male-dominated, her feminine circumvention of the elders may have been punished.

This cultural mishap may have been misconstrued. Perhaps, the locals felt María was making big money providing the 'shroom experience—and that they deserved their share. When they did not get it, they decided to extort her with fire and violence. On our arrival in Huautla, every aspect of the mountain town was about money—pure greed. Another perspective may be that the hippy intruders violated the Mazatec personal space. We traveled on a tight budget and contributed minimally to the indigenous Indian cash flow. Outsiders, focused only on getting high, may have considered the natives as invisible, easy to ignore, or obstacles requiring a detour.

Our journey began only wanting to experience the pre-Columbian cultures that nearby Mexico had to offer. We thought we'd been tuned into, and were part of, the rising counter-cultural movement and we were searching for an adventure in isolated Huautla—Shangri-La. We found the unique, genuine, and exotic. Even though María had welcomed me into her home, I was exactly the type who disrespected the religion of the hongos and basically used María and the Mazatec people only for the experience. I wanted everything for the lowest price. Reading my diary, I can see I was the true ugly American—taking and never giving.

One account reports Richard Nixon had clandestinely authorized the then Mexican President, Luis Echeverria, to harass the hippies. The Mexican Army was supposedly sent to Huautla in 1968 and until 1976 kept the entire area from Teotitlán to Huautla as off-limits to foreigners. We never saw any Army, only municipal authorities with extended, sweaty palms.

Sources infer that María was unhappy with the touristas, but it certainly didn't seem so in my case. María welcomed me—although her family and neighbors tried to squeeze out pesos. Many others profited from the use of hongos and she didn't receive much recompense. We had been told the tale of famous rock stars, actors, and authors visiting María in Huautla, but without facts, it is just folklore. Gordon Wasson made quite a name for himself. To his credit, he never intentionally implicated María or Huautla.

'These young people, blonde and dark-skinned, didn't respect our customs. Never, as far as I remember, were the 'saint children' eaten with such a lack of respect. For me it is not fun to do vigils. Whoever does it simply to feel the effects can go crazy and stay that way temporarily. Our ancestors always took the 'saint children' at a vigil presided over by a Wise One. From the moment the foreigners arrived, 'the saint children' lost their purity. They lost their force, they ruined them. Henceforth they will no longer work. There is no remedy for it.
—María Sabina

Wasson did bring big business to Huautla. He visited many times and was tripped by María on nine different occasions. I wonder how much María received for each visit. A French botanist, Peter Heim, accompanied Wasson's second visit in 1956. Heim identified several species of mushrooms, especially genus Psilocybe and was the first westerner to actually cultivate the magic 'shrooms. Some mushrooms were delivered to Albert Hoffman, who worked for the huge pharmaceutical company Sandoz. In 1938 Hoffman had synthesized LSD. Five years later, he accidently ingested some of his discovery and had a 'psychedelic' experience—followed by more controlled 'experiments'.

With mushrooms from Heim, Hoffman isolated and named psilocybin and psilocin, then published his findings. In 1962, Hoffman travelled to Huautla with Wasson. As a strange twist of fate, they brought synthetic psilocybin for María to try. Sandoz was already marketing the synthetics as Indocybin. María found that a high dose of the pills meant it took longer for her to get off, but they worked about the same.

Another account concerning Wasson, mushrooms, and María is very strange. While Hoffman was synthesizing psilocybin, a man who claimed to be a scientist working at the University of Delaware, James Moore, was also attempting to produce synthetic 'shrooms. Moore was really creating consciousness-altering drugs as futuristic munitions for the CIA! This account includes that Moore tried to buy his way to accompany Wasson on a Huautla expedition with a $2,000 grant.

Everyone made big money off the mushrooms except María. Wasson found fame, sold his books and interviews. Heim sold his cultivated mushrooms. Hoffman and Sandoz made a fortune from synthetic psilocybin. The CIA used the Sandoz-created drug in mind-control experiments. Motivated by Wasson's article, Timothy Leary tried mushrooms unsupervised in Mexico. On his return, Sandoz' synthetic psilocybin was used in altered-states projects at Harvard University. The drug became so popular with the new counter-culture that an estimated one million doses were consumed before it was made illegal in 1966. Everyone got famous and made a fortune. María became internationally famous for all the wrong reasons, but she remained poor.

Because of her fame, María acquired Alvaro Estrada, a local Indian, as her local Boswellian biographer. It is due to Estrada that many of her chants were translated from native Mazatec into Spanish. His brother-in-law, Henry Munn, rendered the Spanish to English. This is one of those translations of María's words and hongos philosophy:

"There is a world beyond ours, a world that is far, also close and invisible. That is where God lives, where the dead and the saints live. A world where everything has already happened; and everything is known. That world speaks. It has its own language. I report what he says. The sacred mushroom takes me by the hand and takes me to the world where everything is known. There are the sacred mushrooms, which speak in a certain way that I can understand. I ask them and they answer me. When I return from the trip I have taken with them, I say what they have told me and what they have shown me."
—María Sabina.

Estrada was also born in Huautla, but he'd learned Spanish in school and directly knew of the mushroom cult. He wrote of María's life in detail. This author also made money—as his now very rare book on María now sells for a thousand dollars a copy.

Nicholas Echeverria produced a film about María's mushroom ritual and chants in 1979. On her deathbed, according to another account, María claimed never to have received any royalties. Her last years were supposedly spent alone. Apparently, a recluse, because she'd been undoubtedly hounded by foreign tourists, like me, who wanted the first-class hongos trip that only María could deliver. She mourned the murder of her son, her family members passing before her, and that the sacred mushrooms, the 'santos niños', had been desecrated.

'I am looking at world peace, but I feel sad. I am looking at people who I knew as a child: my grandmother, my great-grandfather and my parents,' were her last words before she fell into a coma and died.

María Sabina passed from this consciousness, into the clouds she often traveled beyond, on November 22, 1985 at the age of 91. She had been in the hospital for several months being treated for old age. Another of her biographers, Juan Carrero, wrote her local obituary. María was buried, following Indian customs, on Huautla Mountain. She was mourned by her family in a usual wake that lasted three days. Those who met her, will always think well of such a great woman.

María Sabina's Headstone
Here lies the body of a Mazatec woman who, due to her wisdom, was admired by friends and strangers.

The Chicago Tribune headlined María's obituary: Mushroom Drug `Queen` María Sabina. OAXACA, MEXICO — Maria Sabina, an Indian who gained worldwide fame three decades ago as the ``queen of hallucinogenic mushrooms,`` died in her hospital bed Friday. She was 97. She had been treated for nearly a year at Social Security Hospital of Oaxaca, 400 miles southeast of Mexico City, for what doctors called simply old age.

María was never a queen of drugs, just a very expert psychotherapist who used psilocybin to delve into a person's spirit and find the root of the problem. Huautla hasn't changed much over the years. It has a new road, but María is gone. The new sacerdositas now charge a huge fee to trip touristas. María's image is still around mostly of her smoking a cigar everyone confused for a fat joint. Recordings of her ceremonial chants are available: *Mushroom Ceremony of the Mazatec Indians of Mexico*, 1957.
Smithsonian Folkways Recordings

PSILOCYBIN

Mushrooms may have been the first drug experienced by Homo sapiens. María Sabina gave the sacred mushrooms the identity of 'saint children'. The hongos were originally used only for curative purposes. She was the sacred woman, the therapist, the experienced sacerdosita who used hongos therapeutically to cure the soul.

Long before the Europeans brought Christianity to the Western Hemisphere, the Mazatecs were using sacred mushrooms as therapy, combined with religion. The hongos were usually consumed in a group and always with a spiritual leader. Once the 'enlightened' Western World heard of it through the Life Magazine article in 1957, everything concerning the magic hongos changed and changed and changed.

Albert Hoffman, of Sandoz Swiss laboratories, isolated the mind-altering chemicals, as psilocybin, in 1962. This chemical was considered to have the same essence as LSD —which Hoffman had also created in 1938 for treatment of schizophrenia and anxiety.

According to various accounts, Timothy Leary read of Wasson's Life article and drove to Mexico to try the 'shroom —but without spiritual guidance. On his return to Harvard, Leary secured synthetic psilocybin from Sandoz, and with Richard Alpert, began researching its effects to alter some behaviors. They believed that with 'proper' guidance, and in 'proper' settings (not unlike María Sabina's chants) with psychologists, psilocybin could successfully treat alcoholism and reduce the repetition of criminal behaviors.

The Concord Prison Experiment evolved after Leary received synthetic psilocybin from Sandoz. The premise was to witness the effects of psilocybin on paroled convicts. Sixty percent of convicts returned to prison after committing more crimes. After a few doses of 'shroom pills, the control group's recidivism rate dropped to only twenty percent.

This would have been a great break-through if the world had been ready for psychedelic therapy. It wasn't. Due to a myriad of hallucinogenic drugs unleashed on a willing population, in 1966 the government ruled everything illegal. Leary and Alpert were dismissed from Harvard. Leary stayed 'tuned in and turned on'. Alpert dropped out and went to India becoming Ram Dass.

Research on the beneficial effects of psilocybin returned to the headlines in 2016. Psilocybin was used to treat anxiety in cancer patients. Eighty percent of patients had reduced anxiety and were significantly less depressed for up to eight months after the first experience with the drug. Synthetic psilocybin had drastically reduced their fear of death.

Roland Griffiths, a professor of behavioral biology at Johns Hopkins, studied 36 subjects, who had no physical or mental problems and no history of drug use. The study found psilocybin prompted intense spiritual feelings that endured with an optimistic outlook. Other studies are experimenting with psilocybin as treatment for depression in terminally-ill patients —as well as those with anxiety, tobacco addictions, post-traumatic stress, and obsessive-compulsive disorders —with positive results in small clinical trials.

Just as Life Magazine introduced magic mushrooms to a new culture, the New Yorker published an article by Micheal Pollan entitled, The Trip Treatment. The article points again to Johns Hopkins and NYU where cancer patients received one treatment with psilocybin and lost their fear of death.

MycoMediations, a private enterprise in Jamaica, took these studies to fruition and evolved psilocybin use. Seven and ten-day packages are available to enjoy the beach and reflect on life while using psilocybin. One client suffering from cancer reported after taking the mushrooms she not only found peace in this life, but she now looked forward to an afterlife. She went on to say the drug made her feel like she was part of eternity.

The culture that once outlawed psilocybin has turned full circle, realizing what the ancient Mazatecs knew: that everything can be cured from within. María Sabina revealed her tribe's oldest secrets: other dimensions where spirits can be healed. The Mazatec, a small population hidden deep in the mountains of Mexico, altered the world's cultures with the onset of the psychedelic era. In-turn, the world changed the Mazatecs and Huautla. Science finally realizes the sacred use of the magic mushroom was only therapeutic.

María Sabina chanted these verses while in a trance where the *hongos* or *'saint children'* spoke through her:

Because I can swim in the immense
Because I can swim in all forms
Because I am the launch woman
Because I am the sacred opposum
Because I am the Lord opposum

I am the woman Book that is beneath the water, says
I am the woman of the populous town, says
I am the shepherdess who is beneath the water, says
I am the woman who shepherds the immense, says
I am a shepherdess and I come with my shepherd, says

Because everything has its origin
And I come going from place to place from the origin.

Álvaro Estrada and Henry Munn translated many of María's chants.

REFERENCES

The Mex Files: Maria Sabina: Magic mushrooms and silencing the saint children - 14 APRIL 2010
https://mexfiles.net/2010/04/14/maria-sabina/

Life Magazine May 13, 1957, Seeking The Magic Mushroom – Gordon Wasson
https://books.google.com/books?id=Jj8EAAAAMBAJ&pg=PA100&source=gbs_toc_r&cad=2#v=onepage&q&f=false

Seeking The Magic Mushroom
http://www.imaginaria.org/wasson/life.htm
Wikipedia- https://en.wikipedia.org/wiki/Roger_Heim
Wikipedia - https://en.wikipedia.org/wiki/R._Gordon_Wasson
https://en.wikipedia.org/wiki/Mar%C3%ADa_Sabina
Wikipedia - https://en.wikipedia.org/wiki/Huautla_de_Jim%C3%A9nez
Wikipedia - https://en.wikipedia.org/wiki/Mar%C3%ADa_Sabina

The Mazatec Indians - The Mushrooms Speak-
by Henry Munn
http://www.entheology.org/edoto/anmviewer.asp?a=118

Mazateco Newspaper : https://translate.google.com/translate?hl=en&sl=es&u=http://diariomazateco.blogspot.com/2009/10/sabes-porque-huautla-se-llama-de.html&prev=search

Jacket 2 JeromeRothenberg: Henry Munn: From 'The Uniqueness of María Sabina' (In Memoriam)
https://jacket2.org/commentary/henry-munn-uniqueness-mar%C3%ADa-sabina-memoriam

Entheology.com - http://entheology.com/peoples/the-mazatec-indians-the-mushrooms-speak/

Chicago Tribune - http://articles.chicagotribune.com/1985-11-24/news/8503210135_1_hallucinogenic-mushrooms-maria-sabina-indian-custom

HUAUTLA DE JIMÉNEZ OAXACA - https://translate.google.com/translate?hl=en&sl=es&u=https://huautlaoaxaca.wordpress.com/maria-sabina/&prev=search

The Vaults of Erowid - https://erowid.org/culture/characters/wasson_r_gordon/wasson_r_gordon.shtml

The Door of Perception
- http://doorofperception.com/2015/04/r-gordon-wasson-seeking-the-magic-mushroom/

livingandwrokinginmexico - https://livingandworkinginmexico.wordpress.com/2010/08/28/huautla-de-jimenez-in-the-footsteps-of-maria-sabina/

Timeline - This Mexican medicine woman hipped America to magic mushrooms, with the help of a bank executive
- https://timeline.com/with-the-help-of-a-bank-executive-this-mexican-medicine-woman-hipped-america-to-magic-mushrooms-c41f866bbf37

The Devil's Book of Culture: History, Mushrooms, and Caves in Southern Mexico
By Benjamin Feinberg

http://www.stainblue.com/ah.html

Acid Hype: American News Media and the Psychedelic Experience
By Stephen Siff https://books.google.com/books?id=hJr7BwAAQBAJ&pg=PA79&lpg=PA79&dq=Roger+Heim+huautla&source=bl&ots=HZ-1814WABx&sig=5FvrJ2R99dwoEObS8wKSH8XMntE&hl=en&sa=X-&ved=0ahUKEwjfqOj0vqvZAhVJt1MKHXk0D2MQ6AEINDAC#v=onepage&q=Roger%20Heim%20huautla&f=false

CPSIA information can be obtained
at www.ICGtesting.com
Printed in the USA
LVHW080334300122
709722LV00003B/39